SPSS SURVIVAL MANUAL

Second edition

For the SPSS Survival Manual website, go to
www.openup.co.uk/spss

This is what readers from around the world say about the SPSS Survival Manual:

'To any student who have found themselves facing the horror of SPSS after signing up for a degree in psychology—this is a godsend.'

PSYCHOLOGY STUDENT, IRELAND

'This book really lives up to its name . . . I highly recommend this book to any MBA student carrying out a dissertation project, or anyone who needs some basic help with using SPSS and data analysis techniques.'

BUSINESS STUDENT, UK

'If the mere thought of statistics gives you a headache, then this book is for you.'

STATISTICS STUDENT, UK

'. . . one of the most useful, functional pieces of instruction I have seen. So gold stars and thanks.'

INSTRUCTIONAL DESIGNER, USA

'. . . being an external student so much of my time is spent teaching myself. But this has been made easier with your manual as I have found much of the content very easy to follow. I only wish I had discovered it earlier.'

ANTHROPOLOGY STUDENT, AUSTRALIA

'The strength of this book lies in the explanations that accompany the descriptions of tests and I predict great popularity for this text among teachers, lecturers and researchers.'

ROGER WATSON, JOURNAL OF ADVANCED NURSING, 2001

'. . . an excellent book on both using SPSS and statistical know how.'

LECTURER IN BUSINESS RESEARCH METHODS, UK

'SPSS Survival Manual was the only one among loads of SPSS books in the library that was so detailed and easy to follow.'

DOCTORAL STUDENT IN EDUCATION, UK

'My students have sung the book's praises. Teaching statistics, I usually don't get much praise from students for any book.'

STATISTICS LECTURER, USA

'Truly the best SPSS book on the market.'

LECTURER IN MANAGEMENT, AUSTRALIA

'I was behind in class, I was not "getting it" and I was desperate! So I bought all the SPSS books I could find. This book is the one I used. Everything I needed to know and be able to do was clearly explained. The accompanying online database served as an example, showing me how to enter data. This book will not go on my bookshelf; it will remain on my desk through my dissertation and afterwards.'

STUDENT, USA

'This book is exactly what it claims to be— a "survival manual". It contains step by step instructions and clear explanations of how to use SPSS, how to interpret the results, and selecting appropriate tests. This isn't a statistics primer or a text on research design. This is a book for those who haven't had five stats courses and years of using SPSS. If you need help using SPSS to evaluate research data— get this book. A lifesaver!'

STUDENT, USA

'I like it very much and I find it very usefel.'

SOCIOLOGY STUDENT, CZECH REPUBLIC

SPSS SURVIVAL MANUAL

Second edition

A step by step guide to data analysis using
SPSS for Windows (Version 12)

JULIE PALLANT

OPEN UNIVERSITY PRESS

SPPS is a trademark of SPSS Inc. for its proprietary computer software. This book is not approved, sponsored or connected with SPSS Inc. Further information about SPSS is available from:

SPSS Inc., 223 S. Wacker Drive, Chicago, Illinois USA 60606–6307
Tel: 1 (800) 543 2185, Fax: 1 (800) 841 0064
http: //www.spss.com

Open University Press
McGraw-Hill Education
McGraw-Hill House
Shoppenhangers Road
Maidenhead
Berkshire
SL6 2QL
United Kingdom
email: enquiries@openup.co.uk
world wide web: www.openup.co.uk
and
Two Penn Plaza
New York, NY 10121–2289, USA

First edition published in 2001

Copyright © Julie Pallant 2005

Reprinted 2005, 2006

A catalogue record of this book is available from the British Library

ISBN 0 335 21640 4

Library of Congress Cataloging-in-Publication Data available

Set in 10.9/13.68 pt Sabon by Bookhouse, Sydney, Australia
Printed and bound by Bell & Bain Ltd., Glasgow

Contents

Data files and website

Throughout the book you will see examples of research that are taken from a number of data files (*survey.sav, experim.sav*) included on the website that accompanies this book. This website is at:

www.openup.co.uk/spss

To access the data files directly, go to:

www.openup.co.uk/spss/data

From this site you can download the data files to your hard drive or floppy disk by following the instructions on screen. Then you should start SPSS and open the data files. These files can only be opened in SPSS.

The *survey.sav* data file is a 'real' data file, based on a research project that was conducted by one of my graduate diploma classes. So that you can get a feel for the research process from start to finish, I have also included in the Appendix a copy of the questionnaire that was used to generate this data and the codebook used to code the data. This will allow you to follow along with the analyses that are presented in the book, and to experiment further using other variables.

The second data file (*experim.sav*) is a manufactured (fake) data file, constructed and manipulated to illustrate the use of a number of techniques covered in Part Five of the book (e.g. Paired Samples t-test, Repeated Measures ANOVA). This file also includes additional variables that will allow you to practise the skills learnt throughout the book. Just don't get too excited about the results you obtain and attempt to replicate them in your own research!

Two additional data files have been included with this second edition giving you the opportunity to complete some additional activities with data from different discipline areas. The *sleep.sav* file is real datafile from a study conducted to explore the prevalence and impact of sleep problems on aspects of people's lives. The *staffsurvey.sav* file comes from a staff satisfaction survey conducted for a large national educational institution. See the Appendix for further details of these files (and associated materials).

Apart from the data files, the SPSS Survival Manual website also contains a number of useful items for students and instructors, including:

- guidelines for preparing a research report;
- practice exercises;
- updates on changes to SPSS as new versions are released;
- useful links to other websites;

- additional reading; and
- an instructor's guide.

Introduction and overview

This book is designed for students completing research design and statistics courses and for those involved in planning and executing research of their own. Hopefully this guide will give you the confidence to tackle statistical analyses calmly and sensibly, or at least without too much stress!

Many of the problems students experience with statistical analysis are due to anxiety and confusion from dealing with strange jargon, complex underlying theories and too many choices. Unfortunately, most statistics courses and textbooks encourage both of these sensations! In this book I try to translate statistics into a language that can be more easily understood and digested.

The *SPSS Survival Manual* is presented in a very structured format, setting out step by step what you need to do to prepare and analyse your data. Think of your data as the raw ingredients in a recipe. You can choose to cook your 'ingredients' in different ways—a first course, main course, dessert. Depending on what ingredients you have available, different options may, or may not, be suitable. (There is no point planning to make beef stroganoff if all you have is chicken.) Planning and preparation are an important part of the process (both in cooking and in data analysis). Some things you will need to consider are:

- Do you have the correct ingredients in the right amounts?
- What preparation is needed to get the ingredients ready to cook?
- What type of cooking approach will you use (boil, bake, stir-fry)?
- Do you have a picture in your mind of how the end result (e.g. chocolate cake) is supposed to look?
- How will you tell when it is cooked?
- Once it is cooked, how should you serve it so that it looks appetising?

The same questions apply equally well to the process of analysing your data. You must plan your experiment or survey so that it provides the information you need, in the correct format. You must prepare your data file properly and enter your data carefully. You should have a clear idea of your research questions and how you might go about addressing them. You need to know what statistical techniques are available, what sort of data are suitable and what are not. You must be able to perform your chosen statistical technique (e.g. t-test) correctly and interpret the output. Finally, you need to relate this 'output' back to your original research question and know how to present this in your report (or in cooking terms, should you serve your chocolate cake with cream or ice-cream? or perhaps some berries and a sprinkle of icing sugar on top?).

In both cooking and data analysis, you can't just throw in all your ingredients together, shove it in the oven (or SPSS, as the case may be) and pray for the best. Hopefully this book will help you understand the data analysis process a little better and give you the confidence and skills to be a better 'cook'.

Structure of this book

This *SPSS Survival Manual* consists of 22 chapters, covering the research process from designing a study through to the analysis of the data and presentation of the results. It is broken into five main parts. Part One (Getting started) covers the preliminaries: designing a study, preparing a codebook and becoming familiar with SPSS. In Part Two (Preparing the data file) you will be shown how to prepare a data file, enter your data and check for errors. Preliminary analyses are covered in Part Three, which includes chapters on the use of descriptive statistics and graphs; the manipulation of data; and the procedures for checking the reliability of scales. You will also be guided, step by step, through the sometimes difficult task of choosing which statistical technique is suitable for your data.

In Part Four the major statistical techniques that can be used to explore relationships are presented (e.g. correlation, partial correlation, multiple regression, logistic regression and factor analysis). These chapters summarise the purpose of each technique, the underlying assumptions, how to obtain results, how to interpret the output, and how to present these results in your thesis or report.

Part Five discusses the statistical techniques that can be used to compare groups. These include t-tests, analysis of variance, multivariate analysis of variance and analysis of covariance. A chapter on non-parametric techniques is also included.

Using this book

To use this book effectively as a guide to SPSS you need some basic computer skills. In the instructions and examples provided throughout the text I assume that you are already familiar with using a personal computer, particularly the Windows functions. I have listed below some of the skills you will need. Seek help if you have difficulty with any of these operations.

You will need to be able to:

- use the Windows drop-down menus;
- use the left and right buttons on the mouse;

- use the click and drag technique for highlighting text;
- minimise and maximise windows;
- start and exit programs from the Start menu, or Windows Explorer;
- move between programs that are running simultaneously;
- open, save, rename, move and close files;
- work with more than one file at a time, and move between files that are open;
- use Windows Explorer to copy files from the floppy drive to the hard drive, and back again; and
- use Windows Explorer to create folders and to move files between folders.

This book is not designed to 'stand alone'. It is assumed that you have been exposed to the fundamentals of statistics and have access to a statistics text. It is important that you understand some of what goes on 'below the surface' when using SPSS. SPSS is an enormously powerful data analysis package that can handle very complex statistical procedures. This manual does not attempt to cover all the different statistical techniques available in the program. Only the most commonly used statistics are covered. It is designed to get you started and to develop your confidence in using the program.

Depending on your research questions and your data, it may be necessary to tackle some of the more complex analyses available in SPSS. There are many good books available covering the various statistical techniques available with SPSS in more detail. Read as widely as you can. Browse the shelves in your library, look for books that explain statistics in a language that you understand (well, at least some of it anyway!). Collect this material together to form a resource to be used throughout your statistics classes and your research project. It is also useful to collect examples of journal articles where statistical analyses are explained and results are presented. You can use these as models for your final write-up.

The *SPSS Survival Manual* is suitable for use as both an in-class text, where you have an instructor taking you through the various aspects of the research process, and as a self-instruction book for those conducting an individual research project. If you are teaching yourself, be sure to actually practise using SPSS by analysing the data that is included on the website accompanying this book (see p. xi for details). The best way to learn is by actually doing, rather than just reading. 'Play' with the data files from which the examples in the book are taken before you start using your own data file. This will improve your confidence and also allow you to check that you are performing the analyses correctly.

Sometimes you may find that the output you obtain is different from that presented in the book. This is likely to occur if you are using a different version of SPSS to that used throughout this book (SPSS for Windows Version 12). SPSS is updated regularly, which is great in terms of improving the program, but it can lead to confusion for students who find that what is on the screen differs from what is in the book. Usually the difference is not too dramatic, so stay calm and play detective. The information may be there but just in a different form. For information on changes to SPSS for Windows, refer to the website that accompanies this book (see p. xi for details).

Research tips

If you are using this book to guide you through your own research project there are a few additional tips I would like to recommend.

- **Plan your project carefully.** Draw on existing theories and research to guide the design of your project. Know what you are trying to achieve and why.
- **Think ahead.** Anticipate potential problems and hiccups—every project has them! Know what statistics you intend to employ and use this information to guide the formulation of data collection materials. Make sure that you will have the right sort of data to use when you are ready to do your statistical analyses.
- **Get organised.** Keep careful notes of all relevant research, references etc. Work out an effective filing system for the mountain of journal articles you will acquire and, later on, the output from SPSS. It is easy to become disorganised, overwhelmed and confused.
- **Keep good records.** When using SPSS to conduct your analyses, keep careful records of what you do. I recommend to all my students that they buy a spiral bound exercise book to record every session they spend on SPSS. You should record the date, new variables you create, all analyses you perform and also the names of the files where you have saved the SPSS output. If you have a problem, or something goes horribly wrong with your data file, this information can be used by your supervisor to help rescue you!
- **Stay calm!** If this is your first exposure to SPSS and data analysis there may be times when you feel yourself becoming overwhelmed. Take some deep breaths and use some positive self-talk. Just take things step by step—give yourself permission to make mistakes and become confused sometimes. If it all gets too much, then stop, take a walk and clear your head before you tackle it again. Most students find SPSS quite easy to use, once they get the hang of it. Like learning any new skill, you just need to get past that first feeling of confusion and lack of confidence.
- **Give yourself plenty of time.** The research process, particularly the data entry and data analysis stages, always takes longer than expected, so allow plenty of time for this.
- **Work with a friend.** Make use of other students for emotional and practical support during the data analysis process. Social support is a great buffer against stress!

Additional resources

There are a number of different topic areas covered throughout this book, from the initial design of a study, questionnaire construction, basic statistical techniques (t-tests, correlation), through to advanced statistics (multivariate analysis of variance, factor analysis). The References relating to each chapter appear at the end of the chapter. Further reading and resource material can be found in the Recommended References at the end of the book.

Part One

Getting started

Data analysis is only one part of the research process. Before you can use SPSS to analyse your data there are a number of things that need to happen. First, you have to design your study and choose appropriate data collection instruments. Once you have conducted your study, the information obtained must be prepared for entry into SPSS (using something called a 'codebook'). To enter the data into SPSS, you must understand how SPSS works and how to talk to it appropriately. Each of these steps is discussed in Part One. Chapter 1 provides some tips and suggestions for designing a study, with the aim of obtaining good-quality data. Chapter 2 covers the preparation of a codebook to translate the information obtained from your study into a format suitable for SPSS. Finally, in Chapter 3 you are taken on a guided tour of SPSS, and some of the basic skills that you will need are discussed. If this is your first time using SPSS, it is important that you read the material presented in Chapter 3 before attempting any of the analyses presented later in the book.

Part One

Getting started

1 Designing a study

Although it might seem a bit strange to discuss research design in a book on SPSS, it is an essential part of the research process that has implications for the quality of the data collected and analysed. The data you enter into SPSS must come from somewhere—responses to a questionnaire, information collected from interviews, coded observations of actual behaviour, or objective measurements of output or performance. The data are only as good as the instrument that you used to collect them and the research framework that guided their collection.

In this chapter a number of aspects of the research process are discussed that have an impact on the potential quality of the data. First, the overall design of the study is considered; This is followed by a discussion of some of the issues to consider when choosing scales and measures; finally, some guidelines for preparing a questionnaire are presented.

Planning the study

Good research depends on the careful planning and execution of the study. There are many excellent books written on the topic of research design to help you with this process—from a review of the literature, formulation of hypotheses, choice of study design, selection and allocation of subjects, recording of observations and collection of data. Decisions made at each of these stages can affect the quality of the data you have to analyse and the way you address your research questions. In designing your own study I would recommend that you take your time working through the design process to make it the best study that you can produce. Reading a variety of texts on the topic will help. A few good, easy-to-follow titles are Stangor (1998), Goodwin (1998) and, if you are working in the area of market research, Boyce (2003). A good basic overview for health and medical research is Peat (2001).

To get you started, consider these tips when designing your study:

- Consider what type of research design (e.g. experiment, survey, observation) is the best way to address your research question. There are advantages and disadvantages to all types of research approaches; choose the most appropriate approach for your particular research question. Have a good understanding of the research that has already been conducted in your topic area.
- If you choose to use an experiment, decide whether a between-groups design (different subjects in each experimental condition) or a repeated measures

design (same subjects tested under all conditions) is the more appropriate for your research question. There are advantages and disadvantages to each approach (see Stangor, 1998, pp. 176–179), so weigh up each approach carefully.

- In experimental studies make sure you include enough levels in your independent variable. Using only two levels (or groups) means fewer subjects are required, but it limits the conclusions that you can draw. Is a control group necessary or desirable? Will the lack of control group limit the conclusions that you can draw?

- Always select more subjects than you need, particularly if you are using a sample of human subjects. People are notoriously unreliable—they don't turn up when they are supposed to, they get sick, drop out and don't fill out questionnaires properly! So plan accordingly. Err on the side of pessimism rather than optimism.

- In experimental studies, check that you have enough subjects in each of your groups (and try to keep them equal when possible). With small groups it is difficult to detect statistically significant differences between groups (an issue of power, discussed in the introduction to Part Five). There are calculations you can perform to determine the sample size that you will need. See, for example, Stangor (1998, p. 141), or consult other statistical texts under the heading 'power'.

- Wherever possible, randomly assign subjects to each of your experimental conditions, rather than using existing groups. This reduces the problem associated with non-equivalent groups in between-groups designs. Also worth considering is taking additional measurements of the groups to ensure that they don't differ substantially from one another. You may be able to statistically control for differences that you identify (e.g. using analysis of covariance).

- Choose appropriate dependent variables that are valid and reliable (see discussion on this point later in this chapter). It is a good idea to include a number of different measures—some measures are more sensitive than others. Don't put all your eggs in one basket.

- Try to anticipate the possible influence of extraneous or confounding variables. These are variables that could provide an alternative explanation for your results. Sometimes they are hard to spot when you are immersed in designing the study yourself. Always have someone else (supervisor, fellow researcher) check over your design before conducting the study. Do whatever you can to control for these potential confounding variables. Knowing your topic area well can also help you identify possible confounding variables. If there are additional variables that you cannot control, can you measure them? By measuring them, you may be able to control for them statistically (e.g. using analysis of covariance).

- If you are distributing a survey, pilot-test it first to ensure that the instructions, questions, and scale items are clear. Wherever possible, pilot-test on the same type of people who will be used in the main study (e.g. adolescents, unemployed youth, prison inmates). You need to ensure that your respondents can understand the survey or questionnaire items, and respond appropriately. Pilot-testing should also pick up any questions or items that may offend potential respondents.

• If you are conducting an experiment it is a good idea to have a full dress rehearsal and to pilot-test both the experimental manipulation and the dependent measures you intend to use. If you are using equipment, make sure it works properly. If you are using different experimenters or interviewers, make sure they are properly trained and know what to do. If different observers are required to rate behaviours, make sure they know how to appropriately code what they see. Have a practice run and check for inter-rater reliability (how consistent scores are from different raters). Pilot-testing of the procedures and measures helps you identify anything that might go wrong on the day and any additional contaminating factors that might influence the results. Some of these you may not be able to predict (e.g. workers doing noisy construction work just outside the lab's window), but try to control those factors that you can.

Choosing appropriate scales and measures

There are many different ways of collecting 'data', depending on the nature of your research. This might involve measuring output or performance on some objective criteria, or rating behaviour according to a set of specified criteria. It might also involve the use of scales that have been designed to 'operationalise' some underlying construct or attribute that is not directly measurable (e.g. self-esteem).

There are many thousands of validated scales that can be used in research. Finding the right one for your purpose is sometimes difficult. A thorough review of the literature in your topic area is the first place to start. What measures have been used by other researchers in the area? Sometimes the actual items that make up the scales are included in the appendix to a journal article, otherwise you may need to trace back to the original article describing the design and validation of the scale you are interested in. Some scales have been copyrighted, meaning that to use them you need to purchase 'official' copies from the publisher. Other scales, which have been published in their entirety in journal articles, are considered to be 'in the public domain', meaning that they can be used by researchers without charge. It is very important, however, to properly acknowledge each of the scales you use, giving full reference details.

In choosing appropriate scales there are two characteristics that you need to be aware of: reliability and validity. Both of these factors can influence the quality of the data you obtain. When reviewing possible scales to use you should collect information on the reliability and validity of each of the scales. You will need this information for the 'Method' section of your research report. No matter how good the reports are concerning the reliability and validity of your scales, it is important to pilot-test them with your intended sample. Sometimes scales are reliable with some groups (e.g. adults with an English-speaking background), but are totally unreliable when used with other groups (e.g. children from non-English-speaking backgrounds).

Reliability

The reliability of a scale indicates how free it is from random error. Two frequently used indicators of a scale's reliability are test-retest reliability (also referred to as 'temporal stability') and internal consistency. The test-retest reliability of a scale is assessed by administering it to the same people on two different occasions, and calculating the correlation between the two scores obtained. High test-retest correlations indicate a more reliable scale. You need to take into account the nature of the construct that the scale is measuring when considering this type of reliability. A scale designed to measure current mood states is not likely to remain stable over a period of a few weeks. The test-retest reliability of a mood scale, therefore, is likely to be low. You would, however, hope that measures of stable personality characteristics would stay much the same, showing quite high test-retest correlations.

The second aspect of reliability that can be assessed is internal consistency. This is the degree to which the items that make up the scale are all measuring the same underlying attribute (i.e. the extent to which the items 'hang together'). Internal consistency can be measured in a number of ways. The most commonly used statistic is Cronbach's coefficient alpha (available using SPSS, see Chapter 9). This statistic provides an indication of the average correlation among all of the items that make up the scale. Values range from 0 to 1, with higher values indicating greater reliability.

While different levels of reliability are required, depending on the nature and purpose of the scale, Nunnally (1978) recommends a minimum level of .7. Cronbach alpha values are dependent on the number of items in the scale. When there are a small number of items in the scale (fewer than ten), Cronbach alpha values can be quite small. In this situation it may be better to calculate and report the mean inter-item correlation for the items. Optimal mean inter-item correlation values range from .2 to .4 (as recommended by Briggs & Cheek, 1986).

Validity

The validity of a scale refers to the degree to which it measures what it is supposed to measure. Unfortunately, there is no one clear-cut indicator of a scale's validity. The validation of a scale involves the collection of empirical evidence concerning its use. The main types of validity you will see discussed are content validity, criterion validity and construct validity.

Content validity refers to the adequacy with which a measure or scale has sampled from the intended universe or domain of content. *Criterion validity* concerns the relationship between scale scores and some specified, measurable criterion. *Construct validity* involves testing a scale not against a single criterion but in terms of theoretically derived hypotheses concerning the nature of the underlying variable or construct. The construct validity is explored by investigating its relationship with other constructs, both related (convergent validity) and unrelated (discriminant validity). An easy-to-follow summary of the various types of validity is provided in Chapter 5 of Stangor (1998).

There are many good books and articles that can help with the selection of appropriate scales. Some of these are also useful if you need to design a scale yourself. See the References at the end of the chapter.

Preparing a questionnaire

In many studies it is necessary to collect information from your subjects or respondents. This may involve obtaining demographic information from subjects prior to exposing them to some experimental manipulation. Alternatively, it may involve the design of an extensive survey to be distributed to a selected sample of the population. A poorly planned and designed questionnaire will not give good data with which to address your research questions. In preparing a questionnaire, you must consider how you intend to use the information; you must know what statistics you intend to use. Depending on the statistical technique you have in mind, you may need to ask the question in a particular way, or provide different response formats. Some of the factors you need to consider in the design and construction of a questionnaire are outlined in the sections that follow.

This section only briefly skims the surface of the questionnaire design, so I would suggest that you read further on the topic if you are designing your own study. A good book for this purpose is Oppenheim (1992) or if your research area is business, Boyce (2003).

Question types

Most questions can be classified into two groups: closed or open-ended. A closed question involves offering respondents a number of defined response choices. They are asked to mark their response using a tick, cross, circle etc. The choices may be a simple *Yes/No*, *Male/Female*; or may involve a range of different choices, for example:

What is the highest level of education you have completed (please tick)?

1. Primary school ____
2. Some secondary school ____
3. Completed secondary school ____
4. Trade training ____
5. University (undergraduate) ____
6. University (postgraduate) ____

Closed questions are usually quite easy to convert to the numerical format required for SPSS. For example, *Yes* can be coded as a 1, *No* can be coded as a 2; *Males* as 1, *Females* as 2. In the education question shown above, the number corresponding to the response ticked by the respondent would be entered. For example, if the respondent ticked *University (undergraduate)*, then this would be coded as a 5. Numbering each of the possible responses helps with the coding process. For data entry purposes, decide on a convention for the numbering (e.g. in order across the page, and then down), and stick with it throughout the questionnaire.

Sometimes you cannot guess all the possible responses that respondents might make—it is therefore necessary to use open-ended questions. The advantage here is that respondents have the freedom to respond in their own way, not restricted to the choices provided by the researcher.

What is the major source of stress in your life at the moment?

....

....

Responses to open-ended questions can be summarised into a number of different categories for entry into SPSS. These categories are usually identified after looking through the range of responses actually received from the respondents. Some possibilities could also be raised from an understanding of previous research in the area. Each of these response categories is assigned a number (e.g. work=1, finances=2, relationships=3), and this number is entered into SPSS. More details on this are provided in the section on preparing a codebook in Chapter 2.

Sometimes a combination of both closed and open-ended questions works best. This involves providing respondents with a number of defined responses, also an additional category (*other*) that they can tick if the response they wish to give is not listed. A line or two is provided so that they can write the response they wish to give. This combination of closed and open-ended questions is particularly useful in the early stages of research in an area, as it gives an indication of whether the defined response categories adequately cover all the responses that respondents wish to give.

Response format

In asking respondents a question, you also need to decide on a response format. The type of response format you choose can have implications when you come to do your statistical analysis. Some analyses (e.g. correlation) require scores that are continuous, from low through to high, with a wide range of scores. If you had asked respondents to indicate their age by giving them a category to tick (less than 30, between 31 and 50 and over 50), these data would not be suitable to use in a correlational analysis. So, if you intend to explore the correlation between age and, say, self-esteem, you will need to ensure that you ask respondents for their actual age in years.

Try to provide as wide a choice of responses to your questions as possible. You can always condense things later if you need to (see Chapter 8). Don't just ask respondents whether they agree or disagree with a statement—use a Likert-type scale, which can range from strongly disagree to strongly agree:

strongly disagree	1 2 3 4 5 6 7 8 9 10	strongly agree

This type of response scale gives you a wider range of possible scores, and increases the statistical analyses that are available to you. You will need to make a decision concerning the number of response steps (e.g. 1 to 10) you use. DeVellis (1991) has a good discussion concerning the advantages and disadvantages of different response scales.

Whatever type of response format you choose, you must provide clear instructions. Do you want your respondents to tick a box, circle a number, make a mark on a line? For many respondents this may be the first questionnaire that they have completed. Don't assume they know how to respond appropriately. Give clear instructions, provide an example if appropriate, and always pilot-test on the type of people that will make up your sample. Iron out any sources of confusion before distributing hundreds of your questionnaires.

In designing your questions always consider how a respondent might interpret the question and all the possible responses a person might want to make. For example, you may want to know whether people smoke or not. You might ask the question:

Do you smoke? (please circle)   Yes   No

In trialling this questionnaire your respondent might ask, whether you mean cigarettes, cigars or marijuana. Is knowing whether they smoke enough? Should you also find out how much they smoke (two or three cigarettes, versus two or three packs), how often they smoke (every day or only on social occasions)? The message here is to consider each of your questions, what information they will give you and what information might be missing.

Wording the questions

There is a real art to designing clear, well-written questionnaire items. Although there are no clear-cut rules that can guide this process, there are some things you can do to improve the quality of your questions, and therefore your data. Oppenheim (1992) suggests a number of things that you should avoid when formulating your questions. Try to avoid:

- long complex questions;
- double negatives;
- double-barrelled questions;
- jargon or abbreviations;
- culture-specific terms;
- words with double meanings;
- leading questions; and
- emotionally loaded words.

When appropriate, you should consider including a response category for 'Don't know', or 'Not applicable'. For further suggestions on writing questions, see Oppenheim (1992, pp. 128–130).

References

Planning the study
Cooper, D. R., & Schindler, P. S. (2003). *Business research methods* (8th edn). Boston: McGraw-Hill.

Goodwin, C. J. (1998). *Research in psychology: Methods and design* (2nd edn). New York: John Wiley.

Peat, J. (2001) Health science research: A handbook of quantitative methods. Sydney: Allen & Unwin.

Stangor, C. (1998). *Research methods for the behavioral sciences*. Boston: Houghton Mifflin.

Selection of appropriate scales
Briggs, S. R., & Cheek, J. M. (1986). The role of factor analysis in the development and evaluation of personality scales. *Journal of Personality, 54,* 106–148.

Dawis, R. V. (1987). Scale construction. *Journal of Counseling Psychology, 34,* 481–489.

DeVellis, R. F. (1991). *Scale development: Theory and applications*. Newbury, CA: Sage.

Nunnally, J. O. (1978). *Psychometric theory*. New York: McGraw-Hill.

Oppenheim, A. N. (1992). *Questionnaire design, interviewing and attitude measurement*. London: St Martin's Press.

Robinson, J. P., Shaver, P. R., & Wrightsman, L. S. (1991). Criteria for scale selection and evaluation. In J. P. Robinson, P. R. Shaver, & L. S. Wrightsman. (Eds), *Measures of personality and social psychological attitudes* (pp. 1–16). Hillsdale, NJ: Academic Press.

Stangor, C. (1998). *Research methods for the behavioral sciences*. Boston: Houghton Mifflin.

Streiner, D. L. & Norman, G. R. (1995). *Health measurement scales: A practical guide to their development and use* (2nd edn). Oxford: Oxford University Press.

Questionnaire design

Boyce, J. (2003). *Market research in practice*. Boston: McGraw Hill.

De Vellis, R. F. (1991). *Scale development: Theory and applications*. Newbury, CA: Sage.

Oppenheim, A. N. (1992). *Questionnaire design, interviewing and attitude measurement*. London: St Martin's Press.

2 Preparing a codebook

Before you can enter the information from your questionnaire, interviews or experiment into SPSS it is necessary to prepare a 'codebook'. This is a summary of the instructions you will use to convert the information obtained from each subject or case into a format that SPSS can understand. The steps involved will be demonstrated in this chapter using a data file that was developed by a group of my Graduate Diploma students. A copy of the questionnaire, and the codebook that was developed for this questionnaire, can be found in the Appendix. The data file is provided on the website that accompanies this book (see p. xi). The provision of this material allows you to see the whole process, from questionnaire development through to the creation of the final data file ready for analysis. Although I have used a questionnaire to illustrate the steps involved in the development of a codebook, a similar process is also necessary in experimental studies.

Preparing the codebook involves deciding (and documenting) how you will go about:

- defining and labelling each of the variables; and
- assigning numbers to each of the possible responses.

All this information should be recorded in a book or computer file. Keep this somewhere safe; there is nothing worse than coming back to a data file that you haven't used for a while and wondering what the abbreviations and numbers refer to.

In your codebook you should list all of the variables in your questionnaire, the abbreviated variable names that you will use in SPSS and the way in which you will code the responses. In this chapter simplified examples are given to illustrate the various steps.

In the first column of Table 2.1 you have the name of the variable (in English, rather than in computer talk). In the second column you write the abbreviated name for that variable that will appear in SPSS (see conventions below), and in the third column you detail how you will code each of the responses obtained.

Variable names

Each question or item in your questionnaire must have a unique variable name. Some of these names will clearly identify the information (e.g. sex, age). Other questions, such as the items that make up a scale, may be identified using an

Variable	SPSS Variable name	Coding instructions	
			Table 2.1
Identification number	ID	Number assigned to each questionnaire	Example of a codebook
Sex	Sex	1 = Males 2 = Females	
Age	Age	Age in years	
Marital status	Marital	1 = single 2 = steady relationship 3 = married for the first time 4 = remarried 5 = divorced/separated 6 = widowed	
Optimism scale items 1 to 6	op1 to op6	Enter the number circled from 1 (strongly disagree) to 5 (strongly agree)	

abbreviation (e.g. op1, op2, op3 is used to identify the items that make up the Optimism scale).

There are a number of conventions you must follow in assigning names to your variables in SPSS. These are set out in the 'Rules for naming of variables' box. In earlier versions of SPSS (prior to version 12) you could use only 8 characters for your variable names. SPSS version 12 is more generous and allows you 64 characters. If you need to transfer data files between different versions of SPSS (e.g. using university computer labs) it might be safer to set up your file using only 8 character variable names.

Rules for naming of variables

Variable names:

- must be unique (i.e. each variable in a data set must have a different name);
- must begin with a letter (not a number);
- cannot include full stops, blanks or other characters (!, ? * ");
- cannot include words used as commands by SPSS (all, ne, eq, to, le, lt, by, or, gt, and, not, ge, with); and
- cannot exceed 64 characters (for SPSS Version 12) or 8 characters for earlier versions of SPSS.

The first variable in any data set should be ID—that is, a unique number that identifies each case. Before beginning the data entry process, go through and assign a number to each of the questionnaires or data records. Write the number clearly on the front cover. Later, if you find an error in the data set, having the questionnaires or data records numbered allows you to check back and find where the error occurred.

Coding responses

Each response must be assigned a numerical code before it can be entered into SPSS. Some of the information will already be in this format (e.g. age in years), other variables such as sex will need to be converted to numbers (e.g. 1=males, 2=females). If you have used numbers in your questions to label your responses (see, for example, the education question in Chapter 1), this is relatively straightforward. If not, decide on a convention and stick to it. For example, code the first listed response as 1, the second as 2 and so on across the page.

What is your current marital staus? (please tick)

single ____   in a relationship ____   married ____   divorced ____

To code responses to the question above: if a person ticked *single,* they would be coded as 1; if in a *relationship*, they would be coded 2; if *married*, 3; and if *divorced*, 4.

Coding open-ended questions

For open-ended questions (where respondents can provide their own answers), coding is slightly more complicated. Take, for example, the question: *What is the major source of stress in your life at the moment?* To code responses to this you will need to scan through the questionnaires and look for common themes. You might notice a lot of respondents listing their source of stress as related to work, finances, relationships, health or lack of time.

In your codebook you list these major groups of responses under the variable name *stress*, and assign a number to each (work=1, finances=2 and so on). You also need to add another numerical code for responses that did not fall into these listed categories (other=9). When entering the data for each respondent you compare his/her response with those listed in the codebook and enter the appropriate number into the data set under the variable *stress*.

Once you have drawn up your codebook, you are almost ready to enter your data. There are two things you need to do first:

1. get to know SPSS, how to open and close files, become familiar with the various 'windows' and dialogue boxes that it uses.
2. set up a data file, using the information you have prepared in your codebook.

In Chapter 3 the basic structure and conventions of SPSS are covered, followed in Chapter 4 by the procedures needed to set up a data file and to enter data.

3 Getting to know SPSS

There are a few key things to know about SPSS before you start. First, SPSS operates using a number of different screens, or 'windows', designed to do different things. Before you can access these windows you need to either open an existing data file or create one of your own. So, in this chapter, we will cover how to open and close SPSS; how to open and close existing data files; and how to create a data file from scratch. We will then go on to look at the different windows SPSS uses.

Starting SPSS

There are a number of different ways to start SPSS:

- The simplest way is to look for an SPSS icon on your desktop. Place your cursor on the icon and double-click.
- You can also start SPSS by clicking on **Start**, move your cursor up to **Programs**, and then across to the list of programs available. Move up or down until you find **SPSS for Windows**.
- SPSS will also start up if you double-click on an SPSS data file listed in Windows Explorer—these files have a .sav extension.

When you open SPSS you may encounter a grey front cover screen asking 'What would you like to do?'. It is easier to close this screen (click on the cross in the top right-hand corner) and get used to using the other SPSS menus. When you close the opening screen you will see a blank spreadsheet. To open an existing SPSS data file from this spreadsheet screen, click on **File**, and then **Open**, from the menu displayed at the top of the screen.

Opening an existing data file

Hint
If your data file is on a floppy disk it is much faster, easier and safer if you transfer your data file from the A: drive onto the hard drive (usually the C: drive) using Windows Explorer, before starting your SPSS session. Do your data entry or analyses on the hard drive and then, at the end of your session, copy the files back onto your floppy disk. If you are working in a computer lab it may be necessary to check with your lecturer or lab supervisor concerning this process.

If you wish to open an existing data file (e.g. one of the files included on the website that accompanies this book; see p. xi), click on **File** from the menu across the top of the screen, and then choose **Open**, and then **Data**. The **Open File** dialogue box will allow you to search through the various directories on your computer to find where your data file is stored. You should always open data files from the hard drive of your computer, not the Floppy or A: drive. If you have data on a floppy disk, transfer it to a folder on the hard drive of your computer before opening it. Find the file you wish to use and click on **Open**. Remember, all SPSS data files have a .sav extension. The data file will open in front of you in what is labelled the **Data Editor** window (more on this window later).

Working with data files

SPSS will allow you to have only one data file open at any one time. You can, however, change data files during an SPSS session. Although it might seem strange, you don't close one data file and then open another. If you try to, SPSS just closes the whole program down. Instead, you ask SPSS to open a second file and it automatically closes the first one. If you have made any changes to the first file, SPSS will ask if you would like to save the file before closing. If you don't save it, you will lose any data you may have entered and any recoding or computing of new variables you may have done since the file was opened.

Saving a data file

When you first create a data file, or make changes to an existing one (e.g. creating new variables), you must remember to save your data file. This does not happen automatically, as in some word processing programs. If you don't save regularly, and there is a power blackout or you accidentally press the wrong key (it does happen!), you will lose all of your work. So save yourself the heartache and save regularly. If you are entering data, this may need to be as often as every ten minutes or after every five or ten questionnaires.

To save a file you are working on, go to the **File** menu (top left-hand corner) and choose **Save**. Or, if you prefer, you can also click on the icon that looks like a floppy disk, which appears on the toolbar at the top, left of your screen.
Please note: Although this icon looks like a floppy disk, clicking on it will save your file to whichever drive you are currently working on. This should always be the hard drive—working from the A: drive is a recipe for disaster! I have had many students come to me in tears after corrupting their data file by working from the A: drive rather than from the hard disk.

When you first save a new data file, you will be asked to specify a name for the file and to indicate a directory and a folder that it will be stored in. Choose the directory and then type in a file name. SPSS will automatically give all data file names the extension .sav. This is so that it can recognise it as an SPSS data file. Don't change this extension, otherwise SPSS won't be able to find the file when you ask for it again later.

Opening a different data file

If you finish working on a data file and wish to open another one, just click on **File** and then **Open,** and find the directory where your second file is stored. Click on the desired file and then click the **Open** button. This will close the first data file and then open the second. Unlike with a word processor, you cannot close one data file and then open another. You must have a data file open at all times.

Starting a new data file

Starting a new data file is easy in SPSS is easy. Click on **File**, then, from the drop-down menu, click on **New** and then **Data**. From here you can start defining your variables and entering your data. Before you can do this, however, you need to understand a little about the windows and dialogue boxes that SPSS uses. These are discussed in the next section.

SPSS windows

The main windows you will use in SPSS are the **Data Editor**, the **Viewer**, the **Pivot Table Editor, Chart Editor** and the **Syntax Editor**. These windows are summarised here, but are discussed in more detail in later sections of this book.

When you begin to analyse your data you will have a number of these windows open at the same time. Some students find this idea very confusing. Once you get the hang of it, it is really quite simple. You will always have the **Data Editor** open because this contains the data file that you are analysing. Once you start to do some analyses you will have the **Viewer** window open because this is where the results of all your analyses are displayed, listed in the order in which you performed them.

The different windows are like pieces of paper on your desk—you can shuffle them around, so that sometimes one is on top and at other times an other. Each of the windows you have open will be listed along the bottom of your screen. To change windows, just click on whichever window you would like to have on top. You can also click on **Window** on the top menu bar. This will list all the open windows and allow you to choose which you would like to display on the screen.

Sometimes the windows SPSS displays do not initially fill the full screen. It is much easier to have the **Viewer** window (where your results are displayed) enlarged on top, filling the entire screen. To do this, look on the top right-hand area of your screen. There should be three little buttons or icons.

Click on the middle button to maximise that window (i.e. to make your current window fill the screen). If you wish to shrink it down again, just click on this middle icon again.

Data Editor window

The **Data Editor** window displays the contents of your data file, and in this window you can open, save and close existing data files; create a new data file; enter data; make changes to the existing data file; and run statistical analyses (see Figure 3.1).

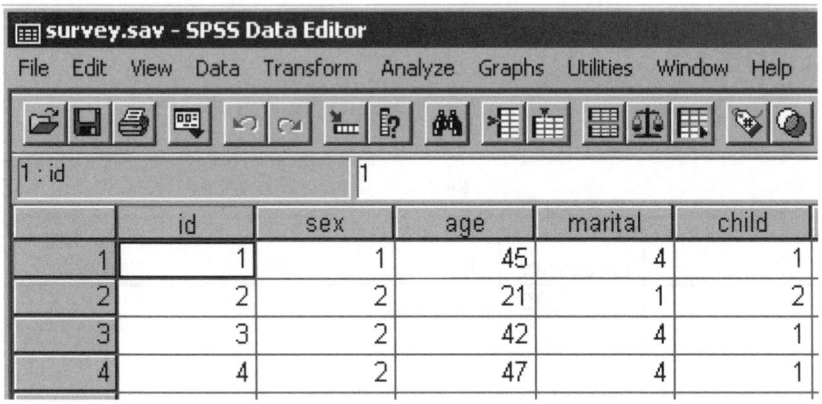

Figure 3.1

Example of a Data Editor window

Viewer window

When you start to do analyses, the **Viewer** window will open automatically (see Figure 3.2). This window displays the results of the analyses you have conducted, including tables and charts. In this window you can modify the output, delete it, copy it, save it, or even transfer it into a Word document. When you save the output from SPSS statistical analyses it is saved in a separate file with a .spo extension, to distinguish it from data files, which have a .sav extension.

The **Viewer** screen consists of two parts. On the left is an outline or menu pane, which gives you a full list of all the analyses you have conducted. You can use this side to quickly navigate your way around your output (which can become very long, very quickly). Just click on the section you want to move to and it will appear on the right-hand side of the screen. On the right-hand side of the **Viewer** window are the results of your analyses, which can include tables and charts (or graphs).

Saving output

To save the results of your analyses you must have the **Viewer** window open on the screen in front of you. Click on **File** from the menu at the top of the screen. Click on **Save**. Choose the directory and folder you wish to save your output in, and then type in a file name that uniquely identifies your output. Click on **Save**. To name my files, I use an abbreviation that indicates the data file I am working on, and the date I conducted the analyses. For example, the file survey8may99.spo would contain the analyses I conducted on 8 May 1999 using the survey data file. I keep a log book that contains a list of all my file names, along with details

Output1 - SPSS Viewer

File   Edit   View   Data   Transform   Insert   Format   Analyze   Graphs   Utilities   Window   Help

☐···· Output
    ☐···· Frequencies
        ···· Title
        ···· Notes
        ►···· sex

Frequencies

sex

		Frequency	Percent	Valid Percent	Cumulative Percent
Valid	MALES	185	42.1	42.1	42.1
	FEMALES	254	57.9	57.9	100.0
	Total	439	100.0	100.0	

Figure 3.2

Example of a Viewer
window

of the analyses that were performed. This makes it much easier for me to retrieve the results of specific analyses. When you begin your own research, you will find that you can very quickly accumulate a lot of different files containing the results of many different analyses. To prevent confusion and frustration, get organised and keep good records of the analyses you have done and of where you have saved the results.

Printing output

You can use the menu pane (left-hand side) of the **Viewer** window to select particular sections of your results to print out. To do this you need to highlight the sections that you want. Click on the first section you want, hold the Ctrl key on your keyboard down and then just click on any other sections you want. To print these sections, click on the File menu (from the top of your screen) and choose **Print**. SPSS will ask whether you want to print your selected output or the whole output.

Pivot Table Editor window

The tables you see in the **Viewer** window (which SPSS calls Pivot Tables) can be modified to suit your needs. To modify a table you need to double-click on it, which takes you into what is known as the **Pivot Table Editor**. You can use this editor to change the look of your table, the size, the fonts used, the dimensions of the columns—you can even swap the presentation of variables around from rows to columns.

Chart Editor window

When you ask SPSS to produce a histogram, bar graph or scatterplot, it initially displays these in the **Viewer** window. If you wish to make changes to the type or presentation of the chart, you need to go into the **Chart Editor** window by double-clicking on your chart. In this window you can modify the appearance and format of your graph, change the fonts, colours, patterns and line markers (see Figure 3.3).

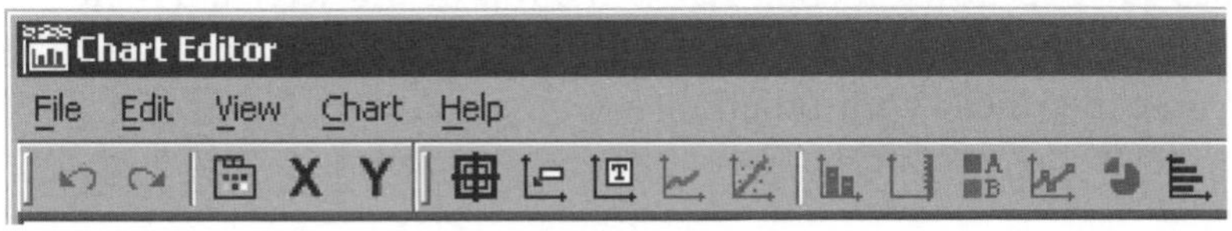

Figure 3.3

Example of a Chart Editor window

    The procedure to generate charts and to use the **Chart Editor** is discussed further in Chapter 7.

Syntax Editor window

In the 'good old days' all SPSS commands were given using a special command language or syntax. SPSS still creates these sets of commands to run each of the programs, but all you usually see are the Windows menus that 'write' the commands for you. Although the options available through the SPSS menus are usually all that most undergraduate students need to use, there are some situations when it is useful to go behind the scenes and to take more control over the analyses that you wish to conduct. This is done using the **Syntax Editor** (see Figure 3.4).

    The **Syntax Editor** is particularly useful when you need to repeat a lot of analyses or generate a number of similar graphs. You can use the normal SPSS

Figure 3.4

Example of a Syntax Editor window

menus to set up the basic commands of a particular statistical technique and then 'paste' these to the **Syntax Editor** (see Figure 3.4). The **Syntax Editor** allows you to copy and paste commands, and to make modifications to the commands generated by SPSS. An example of its use is presented in Chapter 11. Syntax is also a good way of keeping a record of what commands you have used, particularly when you need to do a lot of recoding of variables or computing new variables (demonstrated in Chapter 8).

Menus

Within each of the windows described above, SPSS provides you with quite a bewildering array of menu choices. These choices are displayed using little icons (or pictures), also in drop-down menus across the top of the screen. Try not to become overwhelmed; initially, just learn the key ones, and as you get a bit more confident you can experiment with others.

Dialogue boxes

Once you select a menu option you will usually be asked for further information. This is done in a dialogue box. For example, when you ask SPSS to run **Frequencies** it will display a dialogue box asking you to nominate which variables you want to use (see Figure 3.5).

Figure 3.5

Example of a Frequencies dialogue box

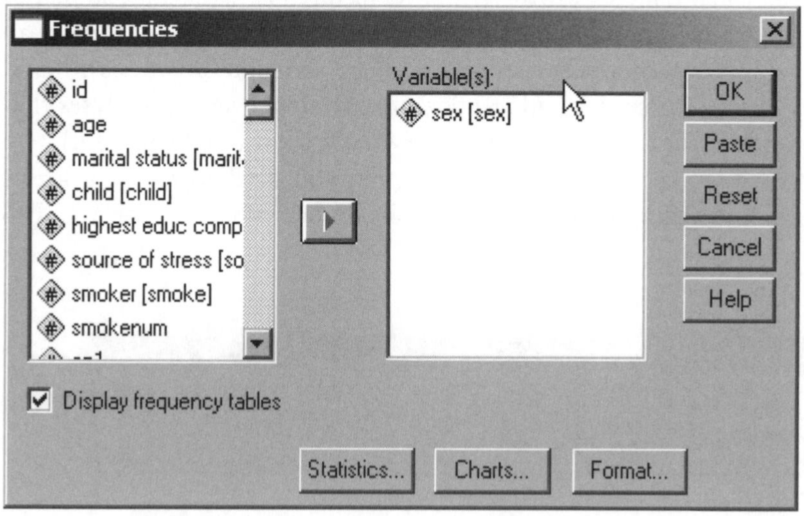

From here, you will open a number of additional sub-dialogue boxes, where you will be able to specify which statistics you would like displayed, the charts that you would like generated and the format the results will be presented in. Different options are available, depending on the procedure or analysis to be performed, but the basic principles in using dialogues boxes are the same. These are discussed below.

Selecting variables in a dialogue box

To indicate which variables you want to use, you need to highlight the selected variables in the list provided (by clicking on them), then click on the arrow button to move them into the empty box labelled **Variable(s)**. To select variables, you can either do this one at a time, clicking on the arrow each time, or you can select a group of variables. If the variables you want to select are all listed together, just click on the first one, hold down the Shift key on your keyboard and press the down arrow key until you have highlighted all the desired variables. Click on the arrow button and all of the selected variables will move across into the **Variable(s)** box.

If the variables you want to select are spread throughout the variable list, you should click on the first variable you want, hold down the Ctrl key, move the cursor down to the next variable you want and then click on it, and so on. Once you have all the desired variables highlighted, click on the arrow button. They will move into the box.

To remove a variable from the box, you just reverse the process. Click on the variable in the **Variable(s)** box that you wish to remove, click on the arrow button, and it shifts the variable back into the original list. You will notice the direction of the arrow button changes, depending on whether you are moving variables into or out of the **Variable(s)** box.

Dialogue box buttons

In most dialogue boxes you will notice a number of standard buttons (**OK, Paste, Reset, Cancel** and **Help;** see Figure 3.5). The uses of each of these buttons are:

- **OK:** click on this button when you have selected your variables and are ready to run the analysis or procedure.
- **Paste:** this button is used to transfer the commands that SPSS has generated in this dialogue box to the **Syntax Editor** (a description of which is presented earlier in this chapter). This is useful if you wish to repeat an analysis a number of times, or if you wish to make changes to the SPSS commands.
- **Reset:** this button is used to clear the dialogue box of all the previous commands you might have given when you last used this particular statistical technique or procedure. It gives you a clean slate to perform a new analysis, with different variables.

- **Cancel:** clicking on this button closes the dialogue box and cancels all of the commands you may have given in relation to that technique or procedure.
- **Help:** click on this button to obtain information about the technique or procedure you are about to perform.

Although I have illustrated the use of dialogue boxes in Figure 3.5 by using **Frequencies,** all dialogue boxes throughout SPSS work on the same basic principle. Each of the dialogue boxes will have a series of buttons with a variety of options relating to the specific procedure or analysis. These buttons will open sub-dialogue boxes that allow you to specify which analyses you wish to conduct or which statistics you would like displayed.

Closing SPSS

When you have finished your SPSS session and wish to close the program down, click on the **File** menu at the top left of the screen. Click on **Exit**. SPSS will prompt you to save your data file and a file that contains your output (results of the analyses). SPSS gives each file an extension to indicate the type of information that it contains. A data file will be given a .sav extension, while the output files will be assigned a .spo extension.

Getting help

If you need help while using SPSS or don't know what some of the options refer to, you can use the in-built Help menu. Click on **Help** from the Menu bar and a number of choices are offered. You can ask for specific topics, work through a **Tutorial,** or consult a **Statistics Coach**. This last choice is an interesting recent addition to SPSS, offering guidance to confused statistics students and researchers. This takes you step by step through the decision-making process involved in choosing the right statistic to use. This is not designed to replace your statistics books, but it may prove a useful guide. The **Results Coach,** also available from the **Help** menu, helps you interpret the output you obtain from some of the statistical procedures.

Within each of the major dialogue boxes there is an additional help menu that will assist you with the procedure you have selected. You can ask about some of the various options that are offered in the sub-dialogue boxes. Move your cursor onto the option you are not sure of and click once with your right mouse button. This brings up a little box that briefly explains the option.

Part Two
Preparing the data file

Preparation of the data file for analysis involves a number of steps. These include creating the data file and entering the information obtained from your study in a format defined by your codebook (covered in Chapter 2). The data file then needs to be checked for errors, and these errors corrected. Part Two of this book covers these two steps. In Chapter 4 the SPSS procedures required to create a data file and enter the data are discussed. In Chapter 5 the process of screening and cleaning the data file is covered.

4 Creating a data file and entering data

In this chapter I will lead you through the process of creating a data file and entering the data using SPSS. There are a number of steps in this process:

- *Step 1*. The first step is to check and modify, where necessary, the options (or preferences, as they were referred to in earlier versions of SPSS) that SPSS uses to display the data and the output that is produced.
- *Step 2*. The next step is to set up the structure of the data file by 'defining' the variables.
- *Step 3*. The final step is to enter the data—that is, the values obtained from each participant or respondent for each variable.

To illustrate these procedures I have used the data file 'survey.sav', which is described in the Appendix. The codebook used to generate these data is also provided in the Appendix.

Data files can also be 'imported' from other spreadsheet-type programs (e.g. Excel). This can make the data entry process much more convenient, particularly for students who don't have SPSS on their home computers. You can set up a basic data file on Excel and enter the data at home. When complete, you can then import the file into SPSS and proceed with the data manipulation and data analysis stages. The instructions for using Excel to enter the data are provided at the end of this chapter.

Changing the SPSS 'Options'

Before you set up your data file it is a good idea to check the SPSS options that govern the way your data and output are displayed. The options allow you to define how your variables will be displayed, the size of your charts, the type of tables that will be displayed in the output and many other aspects of the program. Some of this will seem confusing at first, but once you have used the program to enter data and run some analyses you may want to refer back to this section.

If you are sharing a computer with other people (e.g. in a computer lab), it is worth being aware of these options. Sometimes other students will change these options, which can dramatically influence how the program appears. It is useful to know how to change things back to the way you want them when you come to use the machine.

To open the Options screen, click on **Edit** from the menu at the top of the screen and then choose **Options**. The screen shown in Figure 4.1 should appear. There are a lot of choices listed, many of which you won't need to change. I have described the key ones below, organised by the tab they appear under. To move between the various tabs, just click on the one you want. Don't click on OK until you have finished all the changes you want to make, across all the tabs.

Figure 4.1

Example of Options screen

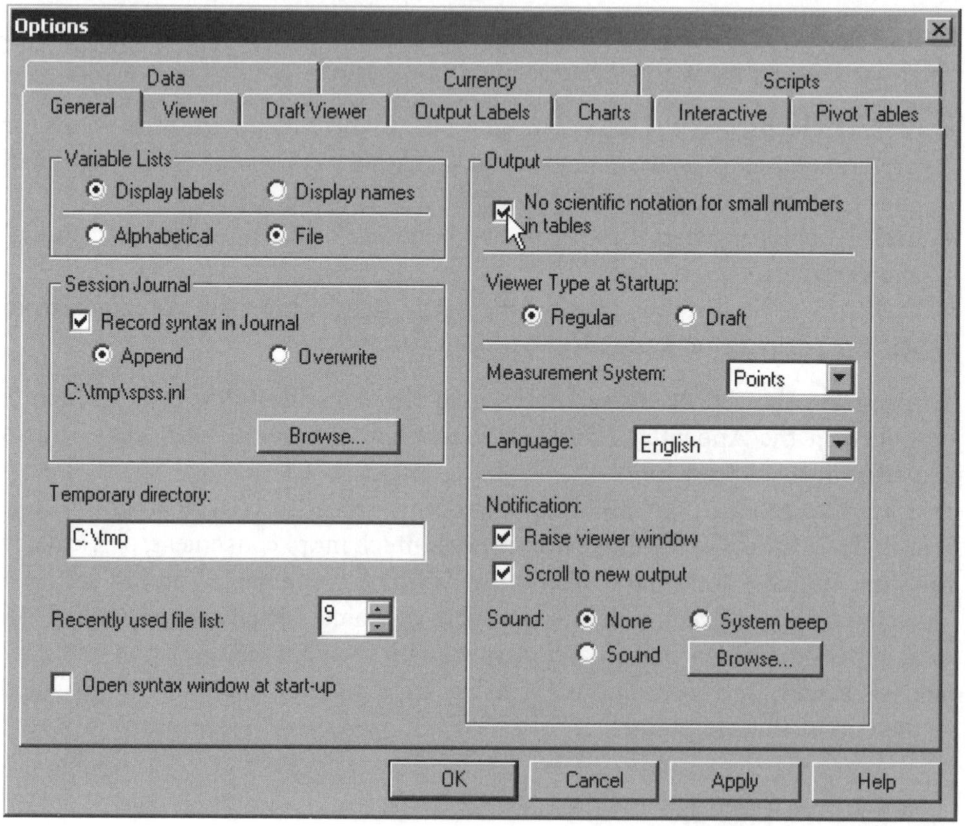

General tab

When you come to do your analyses you can ask for your variables to be listed in alphabetical order, or by the order in which they appear in the file. I always use the file order, because I have all my total scale scores at the end and this keeps them all in a group. Using the file order also means that the variables will remain in the order in which they appear in your codebook. To keep the variables in file order just click on the circle next to **File** in the **Variable Lists** section. In the **Output Notification** section, make sure there is a tick next to **Raise viewer window**, and **Scroll to new output**. This means that when you conduct an analysis the **Viewer** window will appear, and the new output will be displayed on the screen.

In the Output section on the right-hand side, place a tick in the box **No scientific notation for small numbers in tables.** This will stop you getting some very strange numbers in your output for the statistical analyses (see Chapter 13).

Data tab

Click on the Data tab to make changes to the way that your data file is displayed. Make sure there is a tick in the **Calculate values immediately** option. This means that when you calculate a total score the values will be displayed in your data file immediately.

If your variables do not involve values with decimal places, you may like to change the display format for all your variables. In the section labelled **Display format for new numeric variables,** change the decimal place value to 0. This means that all new variables will not display any decimal places. This reduces the size of your data file and simplifies its appearance.

Charts tab

Click on the Charts tab if you wish to change the appearance of your charts. You can alter the **Chart Aspect Ratio** (usually set to 1.5). For some charts a proportion of 1.75 looks better. Experiment with different values to find what suits you. You can also make other changes to the way in which the chart is displayed (e.g. font).

Pivot Tables tab

SPSS presents most of the results of the statistical analyses in tables called Pivot Tables. Under the Pivot Tables tab you can choose the format of these tables from an extensive list. It is a matter of experimenting to find a style that best suits your needs. When I am first doing my analyses I use a style called smallfont.tlo. This saves space (and paper when printing). However, this style is not suitable for importing into documents that need APA style because it includes vertical lines. Styles suitable for APA style are available for when you are ready to format your tables for your research report (see, for example, any of the academic.tlo formats).

You can change the table styles as often as you like—just remember that you have to change the style *before* you run the analysis. You cannot change the style of the tables after they appear in your output, but you can modify many aspects (e.g. font sizes, column width) by using the **Pivot Table Editor.** This can be activated by double-clicking on the table that you wish to modify.

Once you have made all the changes you wish to make on the various options tabs, click on **OK.** You can then proceed to define your variables and enter your data.

Defining the variables

Before you can enter your data, you need to tell SPSS about your variable names and coding instructions. This is called 'defining the variables'. You will do this in the **Data Editor** window (see Figure 4.2). From version 10 of SPSS the **Data Editor** window consists of two different views: **Data View** and **Variable View**. You can move between these two views using the little tabs at the bottom left-hand side of the screen. The **Variable View** is a new SPSS feature, designed to make it easier to define your variables initially and to make changes later as necessary.

You will notice that in the **Data View** window each of the columns is labelled *var* (see Figure 4.2). These will be replaced with the variable names that you listed in your codebook. Down the side you will see the numbers 1, 2, 3 and so on. These are the case numbers that SPSS assigns to each of your lines of data. These are NOT the same as your ID numbers, and these case numbers may change (if, for example, you sort your file or split your file and analyse subsets of your data).

Figure 4.2

Data Editor window

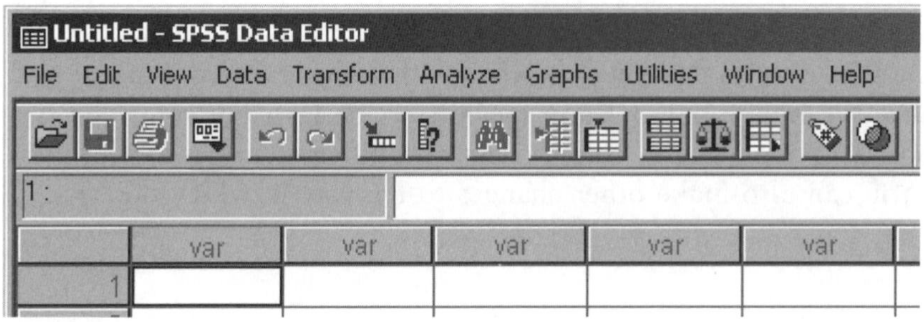

Procedure for defining your variables

To define each of the variables that make up your data file, you first need to click on the **Variable View** tab at the bottom of your screen. In this view (see Figure 4.3) the variables are listed down the side, with their characteristics listed along the top (name, type, width, decimals, label etc.).

Figure 4.3

Variable View

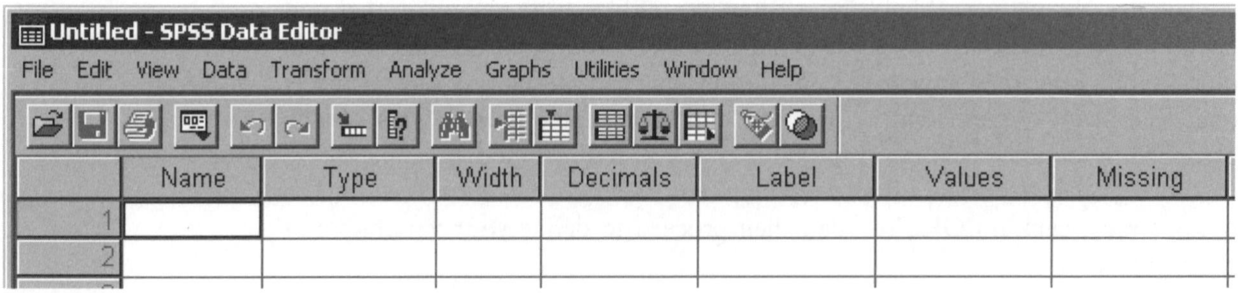

Your job now is to define each of your variables by specifying the required information for each variable listed in your codebook. Some of the information you will need to provide yourself (e.g. name); other bits are provided automatically by SPSS using default values. These default values can be changed if necessary. The key pieces of information that are needed are described below. The headings I have used correspond to the column headings displayed in the **Variable View**. I have provided the simple step-by-step procedures below; however, there are a number of shortcuts that you can use once you are comfortable with the process. These are listed later, in the section headed 'Optional shortcuts'. You should become familiar with the basic techniques first.

Name

In this column, type in the variable name that will be used to identify each of the variables in the data file. These should be listed in your codebook. Each variable name should have only 64 characters (SPSS Version 12) or eight characters or fewer (previous versions of SPSS), and must follow the naming conventions specified by SPSS (these are listed in Chapter 2). Each variable name must be unique. For ideas on how to label your variables, have a look at the codebooks provided in the Appendix. These list the variable names used in the two data files that accompany this book (see p. xi for details of these files).

Type

The default value for Type that will appear automatically as you enter your first variable name is **Numeric**. For most purposes this is all you will need to use. There are some circumstances where other options may be appropriate. If you do need to change this, click on the right-hand side of the cell, where there are three dots. This will display the options available. If your variable can take on values including decimal places, you may also need to adjust the number of decimal places displayed.

Width

The default value for Width is 8. This is usually sufficient for most data. If your variable has very large values you may need to change this default value, otherwise leave it as is.

Decimals

The default value for Decimals (which I have set up using the **Options** facility described earlier in this chapter) is 0. If your variable has decimal places, change this to suit your needs. If all your variables require decimal places, change this under **Options** (using the Data tab). This will save you a lot of time manually changing each of the variables.

Label

The Label column allows you to provide a longer description for your variable than the eight characters that are permitted under the Variable name. This will be used in the output generated from the analyses conducted by SPSS. For example, you may wish to give the label Total Mastery to your variable TMAST.

Values

In the Values column you can define the meaning of the values you have used to code your variables. I will demonstrate this process for the variable 'Sex'.

1. Click on the three dots on the right-hand side of the cell. This opens the **Value Label** dialogue box.
2. Click in the box marked **Value**. Type in *1*.
3. Click in the box marked **Value Label**. Type in *Male*.
4. Click on **Add**. You will then see in the summary box: 1=Male.
5. Repeat for Females: **Value**: enter *2*, **Value Label**: enter *Female*. **Add**.
6. When you have finished defining all the possible values (as listed in your codebook), click on **Continue**.

Missing

Sometimes researchers assign specific values to indicate missing values for their data. This is not essential—SPSS will recognise any blank cell as missing data. So if you intend to leave a blank when a piece of information is not available, it is not necessary to do anything with this Variable View column.

Columns

The default column width is usually set at 8. This is sufficient for most purposes—change it only if necessary to accommodate your values. To make your data file smaller (to fit more on the screen), you may choose to reduce the column width. Just make sure you allow enough space for the width of the variable name.

Align

The alignment of the columns is usually set at 'right' alignment. There is no real need to change this.

Measure

The column heading Measure refers to the level of measurement of each of your variables. The default is **Scale**, which refers to an interval or ratio level of measurement. If your variable consists of categories (e.g. sex), then click in the cell, and then on the arrow key that appears. Choose **Nominal** for categorical data, and **Ordinal** if your data involve rankings, or ordered values.

Optional shortcuts

The process described above can be rather tedious if you have a large number of variables in your data file. There are a number of shortcuts you can use to speed up the process. If you have a number of variables that have the same 'attributes' (e.g. type, width, decimals) you can set the first variable up correctly and then copy these attributes to one or more other variables.

Copying variable definition attributes to one other variable

1. In **Variable View** click on the cell that has the attribute you wish to copy (e.g. Width).

2. From the menu, click on **Edit** and then **Copy**.

3. Click on the same attribute cell for the variable you wish to apply this to.

4. From the menu, click on **Edit** and then **Paste**.

Copying variable definition attributes to a number of other variables

1. In **Variable View** click on the cell that has the attribute you wish to copy (e.g. Width).

2. From the menu, click on **Edit** and then **Copy**.

3. Click on the same attribute cell for the first variable you wish to copy to and then, holding your left mouse button down, drag the cursor down the column to highlight all the variables you wish to copy to.

4. From the menu, click on **Edit** and then **Paste**.

Setting up a series of new variables all with the same attributes

If your data consist of scales made up of a number of individual items, you can create the new variables and define the attributes of all of these items in one go. The procedure is detailed below, using the six items of the Optimism scale as an example (op1 to op6):

1. In **Variable View** define the attributes of the first variable (op1) following the instructions provided earlier. This would involve defining the value labels 1=strongly disagree, 2=disagree, 3=neutral, 4=agree, 5=strongly agree.

2. With the **Variable View** selected, click on the row number of this variable (this should highlight the whole row).

3. From the menu, select **Edit** and then **Copy**.

4. Click on the next empty row.

5. From the menu, select **Edit** and then **Paste Variables**.

6. In the dialogue box that appears enter the number of additional variables you want to add (in this case 5). Enter the prefix you wish to use (op) and the number you wish the new variables to start on (in this case 2). Click on **OK**. This will give you five new variables (op2, op3, op4, op5 and op6).

To set up all of the items in other scales, just repeat the process detailed above (for example, to create the items in the Self-esteem scale I would repeat the same process to define sest1 to sest10). Remember this procedure is suitable only for items that have all the same attributes; it is not appropriate if the items have different response scales (e.g. if some are nominal and others interval level), or if the values are coded differently.

Entering data

Once you have defined each of your variable names and given them value labels (where appropriate), you are ready to enter your data. Make sure you have your codebook ready (see Chapter 2).

Procedure for entering data

1. To enter data you need to have the **Data View** active. Click on the **Data View** tab at the bottom left-hand side of the screen. A spreadsheet should appear with your newly defined variable names listed across the top.

2. Click on the first cell of the data set (first column, first row). A dark border should appear around the active cell.

3. Type in the number (if this variable is *ID* this should be 1, that is case or questionnaire number 1).

4. Press the right arrow key on your keyboard; this will move the cursor into the second cell, ready to enter your second piece of information for case number 1.

5. Move across the row, entering all the information for case 1, making sure that the values are entered in the correct columns.

6. To move back to the start, press the Home key on your keypad. Press the down arrow to move to the second row, and enter the data for case 2.

7. *If you make a mistake and wish to change a value:* Click in the cell that contains the error. The number will appear in the section above the table. Type the correct value in and then press the right arrow key.

After you have defined your variables and entered your data, your Data Editor window should look something like that shown in Figure 4.4 (obviously only a small part of the screen is displayed).

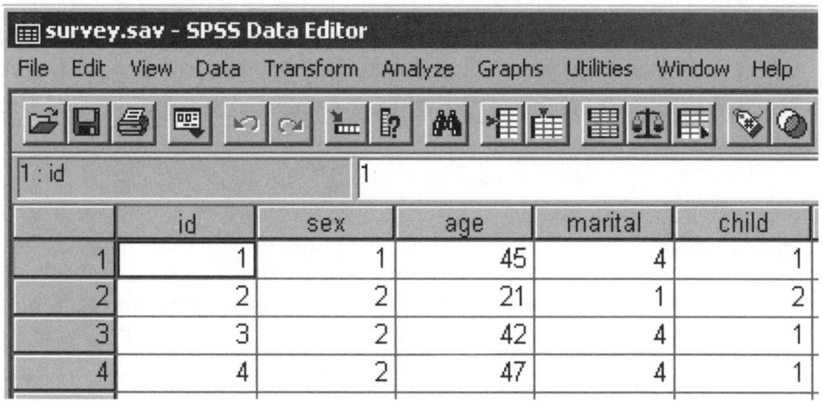

Figure 4.4

Example of a Data Editor window

Modifying the data file

After you have created a data file you may need to make changes to it (e.g. to add, delete or move variables; or to add or delete cases). There are also situations where you may need to sort a data file into a specific order, or to split your file to analyse groups separately. Instructions for each of these actions is given below. Make sure you have the Data Editor window open on the screen.

To delete a case

Move down to the case (row) you wish to delete. Position your cursor in the shaded section on the left-hand side that displays the case number. Click once to highlight the row. Press the Delete button on your computer keypad. You can also click on the **Edit** menu and click on **Clear**.

Important
When entering data, remember to save your data file regularly. SPSS does not automatically save it for you. If you don't save it, you risk losing all the information you have entered. To save, just click on the File menu and choose Save or click on the icon that looks like a computer disk.

To insert a case between existing cases

Move your cursor to a cell in the case (row) immediately below where you would like the new case to appear. Click on the Data menu and choose Insert Case. An empty row will appear in which you can enter the data of the new case.

To delete a variable

Position your cursor in the shaded section (which contains the variable name) above the column you wish to delete. Click once to highlight the whole column. Press the Delete button on your keypad. You can also click on the **Edit** menu and click on **Clear**.

To insert a variable between existing variables

Position your cursor in a cell in the column (variable) to the right of where you would like the new variable to appear. Click on the **Data** menu and choose **Insert Variable**. An empty column will appear in which you can enter the data of the new variable.

To move an existing variable

Create a new empty variable column (follow the previous instructions). Click once on the variable name of the existing variable you wish to move. This should highlight it. Click on the **Edit** menu and choose **Cut**. Highlight the new empty column that you created (click on the name), then click on the **Edit** menu and choose **Paste**. This will insert the variable into its new position.

To sort the data file

You can ask SPSS to sort your data file according to values on one of your variables (e.g. sex, age). Click on the **Data** menu, choose **Sort Cases** and specify which variable will be used to sort by. To return your file to its original order, repeat the process, asking SPSS to sort the file by ID.

To split the data file

Sometimes it is necessary to split your file and to repeat analyses for groups (e.g. males and females) separately. Please note that this procedure does not physically alter your file in any permanent manner. It is an option you can turn on and off as it suits your purposes. The order in which the cases are displayed in the data file will change, however. You can return the data file to its original order (by ID) by using the **Sort Cases** command described above.

To split your file

1. Make sure you have the **Data Editor** window open on the screen.
2. Click on the **Data** menu and choose the **Split File** option.
3. Click on **Compare groups** and specify the grouping variable (e.g. sex).
4. Click on **OK**.

For the analyses that you perform after this split file procedure, the two groups (in this case, males and females) will be analysed separately. When you have finished the analyses, you need to go back and turn the **Split File** option off.

To turn the Split File option off

1. Make sure you have the **Data Editor** window open on the screen.
2. Click on the **Data** menu and choose the **Split File** option.
3. Click on the first dot (**Analyze all cases, do not create groups**).
4. Click on **OK**.

To select cases

For some analyses you may wish to select a subset of your sample (e.g. only males).

To select cases

1. Make sure you have the **Data Editor** window open on the screen.
2. Click on the **Data** menu and choose the **Select Cases** option.
3. Click on the **If condition is satisfied** button.
4. Click on the button labelled **IF . . .**
5. Choose the variable that defines the group that you are interested in (e.g. sex).

6. Click on the arrow button to move the variable name into the box. Click on the = key from the keypad displayed on the screen.

7. Type in the value that corresponds to the group you are interested in (check with your codebook). For example, males in this sample are coded 1, therefore you would type in **1**. The command line should read: sex=1.

8. Click on **Continue** and then **OK**.

For the analyses (e.g. correlation) that you perform after this select cases procedure, only the group that you selected (e.g. males) will be included. When you have finished the analyses, you need to go back and turn the **Select Cases** option off.

To turn the select cases option off

1. Make sure you have the **Data Editor** window open on the screen.

2. Click on the **Data** menu and choose **Select Cases** option.

3. Click on the **All cases option**.

4. Click on **OK**.

Data entry using Excel

Data files can be prepared in the Microsoft Excel program and then imported into SPSS for analysis. This is great for students who don't have access to SPSS at home. Excel usually comes as part of the Microsoft Office package—check under Programs in your Start Menu. The procedure for creating a data file in Excel and then importing it into SPSS is described below. If you intend to use this option, you should have at least a basic understanding of Excel, as this will not be covered here.

One word of warning: Excel can cope with only 256 columns of data (in SPSS language: variables). If your data file is likely to be larger than this, it is probably easier to set it up in SPSS, rather convert from Excel to SPSS later.

Step 1: Set up the variable names

Set up an Excel spreadsheet with the variable names in the first row across the page. The variable names must conform to the SPSS rules for naming variables (see Chapter 2).

Step 2: Enter the data

Enter the information for the first case on one line *across* the page, using the appropriate columns for each variable. Repeat for each of the remaining cases. Don't use any formulas or other Excel functions. Remember to save your file regularly. Click on **File, Save.** In the section marked **Save as Type** make sure 'Microsoft Excel Workbook' is selected. Type in an appropriate file name.

Step 3: Converting to SPSS format

After you have entered the data, save your file and then close Excel. Start SPSS and, with the **Data Editor** open on the screen, click on **File, Open, Data,** from the menu at the top of the screen. In the section labelled **Files of Type** choose Excel. Excel files have a .xls extension. Find the file that contains your data. Click on it so that it appears in the **File name** section. Click on the **Open** button. A screen will appear labelled **Opening Excel Data Source.** Make sure there is a tick in the box: **Read variable names from the first row of data.** Click on **OK.**

The data will appear on the screen with the variable names listed across the top. You will, however, need to go ahead and define the Variable labels, Value labels and the type of Measure. The instructions for these steps are provided earlier in this chapter.

Step 4: Saving as an SPSS file

When you have completed this process of fully defining the variables, you need to save your file as an SPSS file. Choose **File,** and then **Save As** from the menu at the top of the screen. Type in a suitable file name. Make sure that the **Save as Type** is set at SPSS (*.sav). Click on **Save.** When you wish to open this file later to analyse your data using SPSS, make sure you choose the file that has a .sav extension (not your original Excel file that has an .xls extension).

5 Screening and cleaning the data

Before you start to analyse your data it is essential that you check your data set for errors. It is very easy to make mistakes when entering data, and unfortunately some errors can completely mess up your analyses. For example, entering 35 when you mean to enter 3 can distort the results of a correlation analysis. Some analyses are very sensitive to what are known as 'outliers': that is, values that are well below or well above the other scores. So it is important to spend the time checking for mistakes initially, rather than trying to repair the damage later. Although boring, and a threat to your eyesight if you have large data sets, this process is essential and will save you a lot of heartache later!

The data screening process involves a number of steps:

- *Step 1: Checking for errors.* First, you need to check each of your variables for scores that are out of range (i.e. not within the range of possible scores).
- *Step 2: Finding the error in the data file.* Second, you need to find where in the data file this error occurred (i.e. which case is involved).
- *Step 3: Correcting the error in the data file.* Finally, you need to correct the error in the data file itself.

To demonstrate these steps, I have used an example taken from the survey data file (survey.sav) provided on the website accompanying this book (see details on p. xi and in the Appendix). To follow along you will need to start SPSS and open the survey.sav file. This file can be opened only in SPSS. In working through each of the steps on the computer, you will become more familiar with the use of SPSS menus, interpreting the output from SPSS analyses and manipulating your data file.

Step 1: Checking for errors

When checking for errors you are primarily looking for values that fall outside the range of possible values for a variable. For example, if sex is coded 1=male, 2=female, you should not find any scores other than 1 or 2 for this variable.

Scores that fall outside the possible range can distort your statistical analyses—so it is very important that all these errors are corrected before you start.

To check for errors you will need to inspect the frequencies for *each* of your variables. This includes all of the individual items that make up the scales. Errors must be corrected before total scores for these scales are calculated.

There are a number of different ways to check for errors using SPSS. I will illustrate two different ways, one which is more suitable for categorical variables (e.g. sex) and the other for continuous variables (e.g. age). The reason for the difference in the approaches is that some statistics are not appropriate for categorical variables (e.g. it is not appropriate to select **mean** for a variable such as sex with only two values); and with continuous variables you would not want to see a list of all the possible values that the variable can take on.

Checking categorical variables

In this section the procedure for checking categorical variables for errors is presented. In the example shown below I will check the survey.sav data file (included on the website accompanying this book; see p. xi) for errors on the variables Sex, Marital status and Highest education completed.

Procedure for checking categorical variables

1. From the main menu at the top of the screen click on: **Analyze**, then click on **Descriptive Statistics**, then **Frequencies**.
2. Choose the variables that you wish to check (e.g. sex, marital, educ.).
3. Click on the arrow button to move these into the variable box.
4. Click on the **Statistics** button. Tick **Minimum** and **Maximum** in the **Dispersion** section.
5. Click on **Continue** and then on **OK**.

The output generated using this procedure is displayed below (only selected output is displayed).

Statistics

		SEX	Marital status	Highest educ completed
N	Valid	439	439	439
	Missing	0	0	0
Minimum		1	1	1
Maximum		2	8	6

SEX

		Frequency	Percent	Valid Percent	Cumulative Percent
Valid	MALES	185	42.1	42.1	42.1
	FEMALES	254	57.9	57.9	100.0
	Total	439	100.0	100.0	

There are two parts to the output. The first table provides a summary of each of the variables you requested. The remaining tables give you a break-down, for each variable, of the range of responses (these are listed using the value label, rather than the code number that was used).

- Check your minimum and maximum values—do they make sense? Are they within the range of possible scores on that variable? You can see from the first table (labelled **Statistics**) that, for the variable Sex, the minimum value is 1 and the maximum is 2, which is correct. For Marital status the scores range from 1 to 8. Checking this against the codebook, these values are appropriate.
- Check the number of valid cases and missing cases—if there are a lot of missing cases you need to ask why. Have you made errors in entering the data (e.g. put the data in the wrong columns)? Sometimes extra cases appear at the bottom of the data file, where you may have moved your cursor too far down and accidentally created some 'empty' cases. If this occurs, open your **Data Editor** window, move down to the empty case row, click in the shaded area where the case number appears and press Delete on your keypad. Rerun the **Frequencies** procedure again to get the correct values.
- Other tables are also presented in the output, corresponding to each of the variables that were investigated (in this case only selected output on the first variable, sex, is displayed). In these tables you can see how many cases fell into each of the categories (e.g. 185 males, 254 females). Percentages are also presented. This information will be used in the Method section of your report when describing the characteristics of the sample (once any errors have been corrected, of course!).

Checking continuous variables

Procedure for checking continuous variables

1. From the menu at the top of the screen click on **Analyze**, then click on **Descriptive statistics**, then **Descriptives**.

2. Click on the variables that you wish to check. Click on the arrow button to move them into the **Variables** box (e.g. age).

3. Click on the **Options** button. You can ask for a range of statistics, the main ones at this stage are mean, standard deviation, minimum and maximum. Click on the statistics you wish to generate.

4. Click on **Continue**, and then on **OK**.

The output generated from this procedure is shown below.

Descriptive Statistics

	N	Minimum	Maximum	Mean	Std. Deviation
AGE	439	18	82	37.44	13.20
Valid N (listwise)	439				

- Check the minimum and maximum values. Do these make sense? In this case the ages range from 18 to 82.
- Does the mean score make sense? If the variable is the total score on a scale, is the mean value what you expected from previous research on this scale? Is the mean in the middle of the possible range of scores, or is it closer to one end? This sometimes happens when you are measuring constructs such as anxiety or depression.

Step 2: Finding the error in the data file

So what do you do if you find some 'out-of-range' responses (e.g. a 3 for sex). How can you find out where the mistake is in your data set? Don't try to scan through your entire data set looking for the error—there are a number of different ways to find an error in a data file. I will illustrate two approaches.

Procedures for identifying the case where an error has occurred
Method 1

1. Make sure that the **Data Editor** window is open and on the screen in front of you with the data showing.

2. Click on the variable name of the variable in which the error has occurred (e.g. sex).

3. Click once to highlight the column.

4. Click on **Edit** from the menu across the top of the screen. Click on **Find**.

5. In the **Search for** box, type in the incorrect value that you are looking for (e.g. 3).

6. Click on **Search Forward**. SPSS will scan through the file and will stop at the first occurrence of the value that you specified. Take note of the ID number of this case (from the first row). You will need this to check your records or questionnaires to find out what the value should be.

7. Click on **Search Forward** again to continue searching for other cases with the same incorrect value. You may need to do this a number of times before you reach the end of the data set.

Method 2

1. From the menu at the top of the screen click on: **Analyze**, then click on **Descriptive Statistics**, then **Explore**.

2. In the **Display** section click on **Statistics**.

3. Click on the variables that you are interested in (e.g. sex) and move them into the **Dependent list** by clicking on the arrow button.

4. In the **Label cases** section choose **ID** from your variable list. This will give you the ID number of the case, and will allow you to trace back to the questionnaire/record with the mistake.

5. In the **Statistics** section choose **Outliers**. To save unnecessary output you may also like to remove the tick from Descriptives (just click once). Click on **Continue**.

6. In the **Options** section choose **Exclude cases pairwise**. Click on **Continue** and then **OK**.

The output generated from Explore (Method 2) is shown below.

Extreme Values

			Case Number	ID	Value
SEX	Highest	1	3	9	3
		2	209	39	2
		3	241	115	2
		4	356	365	2
		5	345	344	.[a]
	Lowest	1	145	437	1
		2	132	406	1
		3	124	372	1
		4	81	244	1
		5	126	374	.[b]

[a]. Only a partial list of cases with the value 2 are shown in the table of upper extremes.

[b]. Only a partial list of cases with the value 1 are shown in the table of lower extremes.

Note: The data file has been modified for this procedure to illustrate the detection of errors. If you repeat the analyses here using the data files provided on this book's website (see p. xi) you will not find the error as it has been corrected.

Interpretation of output

- The table labelled **Extreme Values** gives you the highest and lowest values recorded for your variable, and also gives you the ID number of the person with that score. Find the value that you know is 'out of range'. In the above example this is a 3.
- Check the ID number given next to the extreme value for that variable. In this printout the person with the ID number of 9 has a value of 3 for Sex. Make sure you refer to the ID number, not the Case number.

Now that we have found which person in the data set has the error, we need to find out what the correct value should be, then go back to the data set to correct it.

Step 3: Correcting the error in the data file

There are a number of steps in this process of correcting an error in the data file.

Procedure for correcting the error in the data file

1. To correct the error, it will be necessary to go back to your questionnaires (or the records from your experiment). Find the questionnaire or record with the ID number that was identified as an extreme value. Check what value should have been entered for that person (e.g. for sex: was it a male (score 1) or a female (score 2)?).

2. Open the **Data Editor** window if it is not already open in front of you. To do this, click on **Window** from the top menu bar, and then on **SPSS Data Editor**.

3. In the data file, find the variable (column) labelled ID. It should be the first one.

4. Move down to the case that has the ID number with the error. Remember that you must use the variable ID column, not the case number on the side of the screen.

5. Once you have found the person with that ID number, move across the row until you come to the column of the variable with the error (e.g. Sex). Place the cursor in the cell, make sure that it is highlighted and then just type in the correct value.

This will replace the old incorrect value. Press one of the arrow keys and you will see the correct value appear in the cell.

After you have corrected your errors it is a good idea to repeat **Frequencies** to double-check. Sometimes, in correcting one error, you will have accidentally caused another error. Although this process is tedious it is very important that you start with a clean, error-free data set. The success of your research depends on it! Don't cut corners.

Reference

For additional information on the screening and cleaning process, I would strongly recommend you read Chapter 4 in:
Tabachnick, B. G., & Fidell, L. S. (2001). *Using multivariate statistics* (4th edn). New York: HarperCollins.

Part Three

Preliminary analyses

Once you have a clean data file, you can begin the process of inspecting your data file and exploring the nature of your variables. This is in readiness for conducting specific statistical techniques to address your research questions. There are five chapters that make up Part Three of this book. In Chapter 6 the procedures required to obtain descriptive statistics for both categorical and continuous variables are presented. This chapter also covers checking the distribution of scores on continuous variables in terms of normality and possible outliers.

Graphs can be useful tools when getting to know your data. Some of the more commonly used graphs available through SPSS are presented in Chapter 7. Sometimes manipulation of the data file is needed to make it suitable for specific analyses. This may involve calculating the total score on a scale, by adding up the scores obtained on each of the individual items. It may also involve collapsing a continuous variable into a smaller number of discrete categories. These data manipulation techniques are covered in Chapter 8. In Chapter 9 the procedure used to check the reliability (internal consistency) of a scale is presented. This is particularly important in survey research, or in studies that involve the use of scales to measure personality characteristics, attitudes, beliefs etc.

Also included in Part Three is a chapter that helps you through the decision-making process in deciding which statistical technique is suitable to address your research question. In Chapter 10 you are provided with an overview of some of the statistical techniques available in SPSS and led step by step through the process of deciding which one would suit your needs. Important aspects that you need to consider (e.g. type of question, data type, characteristics of the variables) are highlighted.

6 Descriptive statistics

Once you are sure there are no errors in the data file (or at least no out-of-range values on any of the variables), you can begin the descriptive phase of your data analysis. Descriptive statistics have a number of uses. These include:

- to describe the characteristics of your sample in the Method section of your report;
- to check your variables for any violation of the assumptions underlying the statistical techniques that you will use to address your research questions; and
- to address specific research questions.

The two procedures outlined in Chapter 5 for checking the data will also give you information for describing your sample in the Method section of your report. In studies involving human subjects, it is useful to collect information on the number of people or cases in the sample, the number and percentage of males and females in the sample, the range and mean of ages, education level, and any other relevant background information.

Prior to doing many of the statistical analyses (e.g. t-test, ANOVA, correlation) it is important to check that you are not violating any of the 'assumptions' made by the individual tests (these are covered in detail in Part Four and Part Five of this book). Testing of assumptions usually involves obtaining descriptive statistics on your variables. These descriptive statistics include the mean, standard deviation, range of scores, skewness and kurtosis.

Descriptive statistics can be obtained a number of different ways, using **Frequencies, Descriptives** or **Explore**. These are all procedures listed under the **Analyze, Descriptive Statistics** drop-down menu. There are, however, different procedures depending on whether you have a categorical or continuous variable. Some of the statistics (e.g. mean, standard deviation) are not appropriate if you have a categorical variable. The different approaches to be used with categorical and continuous variables are presented in the following two sections.

Categorical variables

To obtain descriptive statistics for *categorical* variables you should use **Frequencies**. This will tell you how many people gave each response (e.g. how many males, how many females). It doesn't make any sense asking for means, standard deviations etc. for categorical variables, such as sex or marital status.

Procedure for obtaining descriptive statistics for categorical variables

1. From the menu at the top of the screen click on: **Analyze**, then click on **Descriptive Statistics**, then **Frequencies**.

2. Choose and highlight the categorical variables you are interested in (e.g. sex). Move these into the **Variables** box.

3. Click on the **Statistics** button. In the **Dispersion** section tick **Minimum** and **Maximum**. Click on **Continue** and then **OK**.

The output generated from this procedure is shown below.

Statistics

SEX

N	Valid	439
	Missing	0
Minimum		1
Maximum		2

SEX

		Frequency	Percent	Valid Percent	Cumulative Percent
Valid	MALES	185	42.1	42.1	42.1
	FEMALES	254	57.9	57.9	100.0
	Total	439	100.0	100.0	

Interpretation of output from frequencies

From the output shown above we know that there are 185 males (42.1 per cent) and 254 females (57.9 per cent) in the sample, giving a total of 439 respondents. It is important to take note of the number of respondents you have in different subgroups in your sample. For some analyses (e.g. ANOVA) it is easier to have roughly equal group sizes. If you have very unequal group sizes, particularly if the group sizes are small, it may be inappropriate to run some analyses.

Continuous variables

For continuous variables (e.g. age) it is easier to use **Descriptives,** which will provide you with 'summary' statistics such as mean, median, standard deviation. You certainly don't want every single value listed, as this may involve hundreds of values for some variables. You can collect the descriptive information on all your continuous variables in one go; it is not necessary to do it variable by variable. Just transfer all the variables you are interested in into the box labelled **Variables**. If you have a lot of variables, however, your output will be extremely long. Sometimes it is easier to do them in chunks and tick off each group of variables as you do them.

Procedure for obtaining descriptive statistics for continuous variables

1. From the menu at the top of the screen click on: **Analyze**, then click on **Descriptive Statistics**, then **Descriptives**.

2. Click on all the continuous variables that you wish to obtain descriptive statistics for. Click on the arrow button to move them into the **Variables** box (e.g. age, total perceived stress etc.).

3. Click on the **Options** button. Click on **mean**, **standard deviation**, **minimum**, **maximum**, **skewness**, **kurtosis**.

4. Click on **Continue**, and then **OK**.

The output generated from this procedure is shown below.

Descriptive Statistics

	N	Minimum	Maximum	Mean	Std.	Skewness		Kurtosis	
	Statistic	Statistic	Statistic	Statistic	Statistic	Statistic	Std. Error	Statistic	Std. Error
AGE	439	18	82	37.44	13.20	.606	.117	-.203	.233
Total perceived stress	433	12	46	26.73	5.85	.245	.117	.182	.234
Total Optimism	435	7	30	22.12	4.43	-.494	.117	.214	.234
Total Mastery	436	8	28	21.76	3.97	-.613	.117	.285	.233
Total PCOISS	431	20	88	60.60	11.99	-.395	.118	.247	.235
Valid N (listwise)	425								

Interpretation of output from descriptives

In the output presented above the information we requested for each of the variables is summarised. For example, concerning the variable age, we have information from 439 respondents, the range of ages is from 18 to 82 years, with a mean of 37.44 and standard deviation of 13.20. This information may be needed for the Method section of a report to describe the characteristics of the sample.

Descriptives also provides some information concerning the distribution of scores on continuous variables (skewness and kurtosis). This information may be needed if these variables are to be used in parametric statistical techniques (e.g. t-tests, analysis of variance). The skewness value provides an indication of the symmetry of the distribution. Kurtosis, on the other hand, provides information about the 'peakedness' of the distribution. If the distribution is perfectly normal you would obtain a skewness and kurtosis value of 0 (rather an uncommon occurrence in the social sciences).

Positive skewness values indicate positive skew (scores clustered to the left at the low values). Negative skewness values indicate a clustering of scores at the high end (right-hand side of a graph). Positive kurtosis values indicate that the distribution is rather peaked (clustered in the centre), with long thin tails. Kurtosis values below 0 indicate a distribution that is relatively flat (too many cases in the extremes). With reasonably large samples, skewness will not 'make a substantive difference in the analysis' (Tabachnick & Fidell, 2001, p. 74). Kurtosis can result in an underestimate of the variance, but this risk is also reduced with a large sample (200+ cases: see Tabachnick & Fidell, 2001, p. 75).

While there are tests that you can use to evaluate skewness and kurtosis values, these are too sensitive with large samples. Tabachnick and Fidell (2001, p. 73) recommend inspecting the shape of the distribution (e.g. using a histogram). The procedure for further assessing the normality of the distribution of scores is provided later in this section.

Missing data

When you are doing research, particularly with human beings, it is very rare that you will obtain complete data from every case. It is important that you inspect your data file for missing data. Run **Descriptives** and find out what percentage of values is missing for each of your variables. If you find a variable with a lot of unexpected missing data you need to ask yourself why. You should also consider whether your missing values are happening randomly, or whether there is some systematic pattern (e.g. lots of women failing to answer the question about their age). SPSS has a **Missing Value Analysis** procedure which may help to find patterns in your missing values (see the bottom option under the **Analyze** menu).

You also need to consider how you will deal with missing values when you come to do your statistical analyses. The **Options** button in many of the SPSS statistical procedures offers you choices for how you want SPSS to deal with missing data. It is important that you choose carefully, as it can have dramatic effects on your results. This is particularly important if you are including a list of variables, and repeating the same analysis for all variables (e.g. correlations among a group of variables, t-tests for a series of dependent variables).

- The *Exclude cases listwise* option will include cases in the analysis only if it has full data on *all of the variables* listed in your variables box for that case. A case will be totally excluded from all the analyses if it is missing even one piece of information. This can severely, and unnecessarily, limit your sample size.
- The *Exclude cases pairwise* option, however, excludes the case (person) only if they are missing the data required for the specific analysis. They will still be included in any of the analyses for which they have the necessary information.

- The *Replace with mean* option, which is available in some SPSS statistical procedures (e.g. multiple regression), calculates the mean value for the variable and gives every missing case this value. This option should NEVER be used, as it can severely distort the results of your analysis, particularly if you have a lot of missing values.

Always press the **Options** button for any statistical procedure you conduct, and check which of these options is ticked (the default option varies across procedures). I would strongly recommend that you use pairwise exclusion of missing data, unless you have a pressing reason to do otherwise. The only situation where you might need to use listwise exclusion is when you want to refer only to a subset of cases that provided a full set of results.

Assessing normality

Many of the statistical techniques presented in Part Four and Part Five of this book assume that the distribution of scores on the dependent variable is 'normal'. Normal is used to describe a symmetrical, bell-shaped curve, which has the greatest frequency of scores in the middle, with smaller frequencies towards the extremes (see Gravetter & Wallnau, 2000, p. 52). Normality can be assessed to some extent by obtaining skewness and kurtosis values (as described in the previous section). However, other techniques are also available in SPSS using the **Explore** option of the **Descriptive Statistics** menu. This procedure is detailed below.

In this example I will assess the normality of the distribution of scores for Total Perceived Stress. I have done this separately for males and females (using the **Factor List** option that is available in the **Explore** dialogue box). This is the procedure you would use in preparation for a t-test to explore the difference in Total Perceived Stress scores for males and females (see t-tests for independent samples in Chapter 16). If you wish to assess the normality for the sample as a whole, you will just ignore the instructions given below concerning the Factor List.

Procedure for assessing normality using Explore

1. From the menu at the top of the screen click on: **Analyze**, then click on **Descriptive Statistics**, then **Explore**.
2. Click on the variable/s you are interested in (e.g. total perceived stress). Click on the arrow button to move them into the Dependent List box.
3. Click on any independent or grouping variables that you wish to split your sample by (e.g. sex). Click on the arrow button to move them into the **Factor List** box.

4. In the **Display** section make sure that **Both** is selected. This displays both the plots and statistics generated.
5. Click on the **Plots** button. Under **Descriptive** click on the **Histogram**. Click on **Normality plots with tests**.
6. Click on **Continue**.
7. Click on the **Options** button. In the **Missing Values** section click on **Exclude cases pairwise**.
8. Click on **Continue** and then **OK**.

The output generated from this procedure is shown below.

Descriptives

SEX				Statistic	Std. Error
Total perceived stress	MALES	Mean		25.79	.40
		95% Confidence Interval for Mean	Lower Bound	25.00	
			Upper Bound	26.58	
		5% Trimmed Mean		25.74	
		Median		25.00	
		Variance		29.315	
		Std. Deviation		5.41	
		Minimum		13	
		Maximum		46	
		Range		33	
		Interquartile Range		8.00	
		Skewness		.271	.179
		Kurtosis		.393	.356
	FEMALES	Mean		27.42	.38
		95% Confidence Interval for Mean	Lower Bound	26.66	
			Upper Bound	28.18	
		5% Trimmed Mean		27.35	
		Median		27.00	
		Variance		36.793	
		Std. Deviation		6.07	
		Minimum		12	
		Maximum		44	
		Range		32	
		Interquartile Range		7.00	
		Skewness		.173	.154
		Kurtosis		.074	.307

Tests of Normality

	sex	Kolmogorov-Smirnov[a]			Shapiro-Wilk		
		Statistic	df	Sig.	Statistic	df	Sig.
total perceived stress	MALES	.074	184	.015	.987	184	.096
	FEMALES	.064	249	.015	.992	249	.176

a. Lilliefors Significance Correction

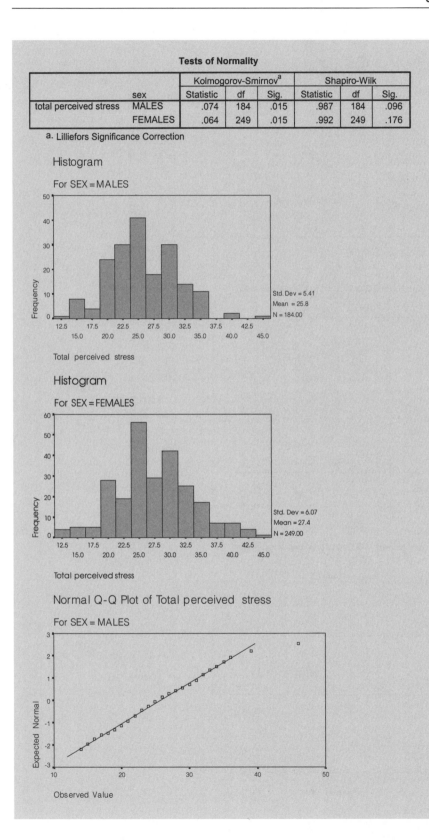

Histogram

For SEX = MALES

Std. Dev = 5.41
Mean = 25.8
N = 184.00

Total perceived stress

Histogram

For SEX = FEMALES

Std. Dev = 6.07
Mean = 27.4
N = 249.00

Total perceived stress

Normal Q-Q Plot of Total perceived stress

For SEX = MALES

Observed Value

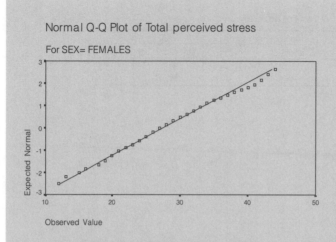

Normal Q-Q Plot of Total perceived stress

For SEX= FEMALES

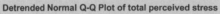

Detrended Normal Q-Q Plot of total perceived stress

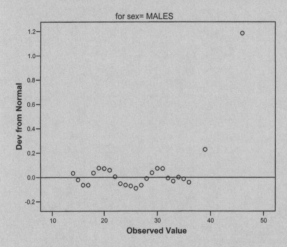

for sex= MALES

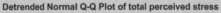

Detrended Normal Q-Q Plot of total perceived stress

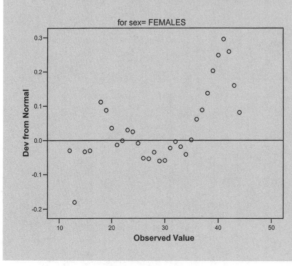

for sex= FEMALES

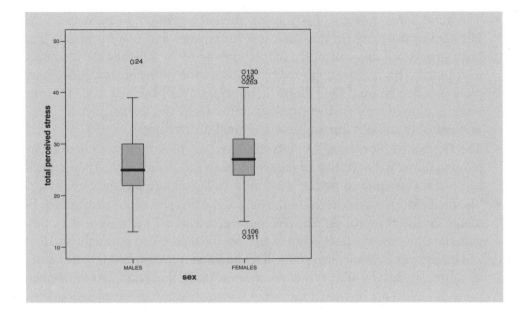

Interpretation of output from explore

Quite a lot of information is generated as part of this output. This tends to be a bit overwhelming until you know what to look for. In this section I will take you through the output step by step.

- In the table labelled **Descriptives** you are provided with descriptive statistics and other information concerning your variables. In this case the table has been divided into two sections corresponding to the two groups, males and females. If you did not specify a grouping variable in the Factor List this information will be provided for the sample as a whole. Some of this information you will recognise (mean, median, std deviation, minimum, maximum etc.). One statistic you may not know is the 5% Trimmed Mean. To obtain this value SPSS removes the top and bottom 5 per cent of your cases and recalculates a new mean value. If you compare the original mean and this new trimmed mean you can see whether some of your more extreme scores are having a strong influence on the mean. If these two mean values are very different, you may need to investigate these data points further.
- Skewness and kurtosis values are also provided as part of this output, giving information about the distribution of scores for the two groups (see discussion of the meaning of these values in the previous section).
- In the table labelled **Tests of Normality** you are given the results of the Kolmogorov-Smirnov statistic. This assesses the normality of the distribution of scores. A non-significant result (Sig value of more than .05) indicates normality. In this case the Sig. value is .015 for each group, suggesting violation of the assumption of normality. This is quite common in larger samples.

- The actual shape of the distribution for each group can be seen in the **Histograms** provided (in this case, one for females and one for males). For both groups in this example, scores appear to be reasonably normally distributed. This is also supported by an inspection of the normal probability plots (labelled **Normal Q-Q Plots**). In these plots the observed value for each score is plotted against the expected value from the normal distribution. A reasonably straight line suggests a normal distribution.
- The **Detrended Normal Q-Q Plots** displayed in the output are obtained by plotting the actual deviation of the scores from the straight line. There should be no real clustering of points, with most collecting around the zero line.
- The final plot that is provided in the output is a boxplot of the distribution of scores for the two groups. The rectangle represents 50 per cent of the cases, with the whiskers (the lines protruding from the box) going out to the smallest and largest values. Sometimes you will see additional circles outside this range— these are classified by SPSS as outliers. The line inside the rectangle is the median value. Boxplots are discussed further in the next section on detecting outliers.

In the example given above, the distribution of scores for both groups was reasonably 'normal'. Often this is not the case. Many scales and measures used in the social sciences have scores that are skewed, either positively or negatively. This does not necessarily indicate a problem with the scale, but rather reflects the underlying nature of the construct being measured. Life satisfaction measures, for example, are often negatively skewed, with most people being reasonably happy with their lot in life. Clinical measures of anxiety or depression are often positively skewed in the general population, with most people recording relatively few symptoms of these disorders. Some authors in this area recommend that, with skewed data, the scores be 'transformed' statistically. This issue is discussed further in Chapter 8 of this book.

Checking for outliers

Many of the statistical techniques covered in this book are sensitive to outliers (cases with values well above or well below the majority of other cases). The techniques described in the previous section can also be used to check for outliers, but an additional approach is detailed below. You will recognise it from Chapter 5, when it was used to check for out-of-range cases.

Procedure for identifying outliers

1. From the menu at the top of the screen click on: **Analyze**, then click on **Descriptive Statistics**, then **Explore**.

2. In the **Display** section make sure **Both** is selected. This provides both Statistics and Plots.

3. Click on your variable (e.g. total perceived stress), and move it into the **Dependent list** box.

4. Click on **id** from your variable list and move into the section **Label cases**. This will give you the ID number of the outlying case.

5. Click on the **Statistics** button. Click on **Outliers**. Click on **Continue**.

6. Click on the **Plots** button. Click on **Histogram**. You can also ask for a Stem and Leaf plot as well if you wish.

7. Click on the **Options** button. Click on **Exclude cases pairwise**. Click on **Continue** and then **OK**.

The output generated from this procedure is shown below.

Descriptives

			Statistic	Std. Error
Total perceived stress	Mean		26.73	.28
	95% Confidence Interval for Mean	Lower Bound	26.18	
		Upper Bound	27.28	
	5% Trimmed Mean		26.64	
	Median		26.00	
	Variance		34.194	
	Std. Deviation		5.85	
	Minimum		12	
	Maximum		46	
	Range		34	
	Interquartile Range		8.00	
	Skewness		.245	.117
	Kurtosis		.182	.234

Extreme Values

			Case Number	ID	Value
Total perceived stress	Highest	1	24	24	46
		2	130	157	44
		3	55	61	43
		4	123	144	42
		5	263	330	.ᵃ
	Lowest	1	5	5	12
		2	311	404	12
		3	103	119	13
		4	239	301	13
		5	106	127	13

a. Only a partial list of cases with the value 42 is shown in the table of upper extremes.

Histogram

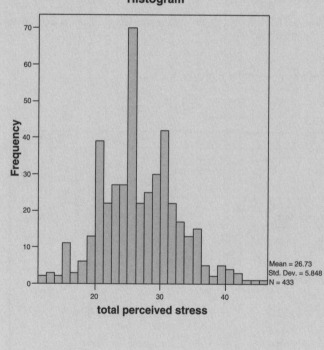

Mean = 26.73
Std. Dev. = 5.848
N = 433

total perceived stress

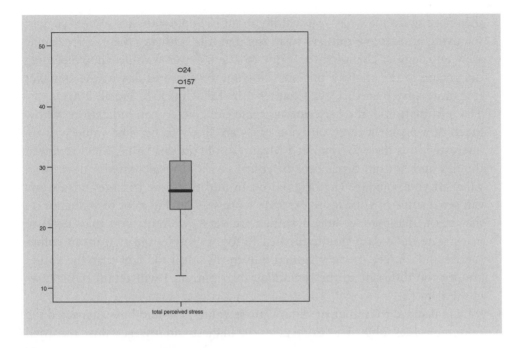

Interpretation of output from explore

This output gives you a number of pieces of useful information concerning the distribution of scores on your variable.

- First, have a look at the Histogram. Look at the tails of the distribution. Are there data points sitting on their own, out on the extremes? If so, these are potential outliers. If the scores drop away in a reasonably even slope then there is probably not too much to worry about.
- Second, inspect the **Boxplot**. Any scores that SPSS considers are outliers appear as little circles with a number attached (this is the ID number of the case). SPSS defines points as outliers if they extend more than 1.5 box-lengths from the edge of the box. Extreme points (indicated with an asterisk *) are those that extend more than 3 box-lengths from the edge of the box. In the example above there are no extreme points, but there are three outliers: ID numbers 24, 157 and 61. If you find points like this you need to decide what to do with them.
- It is important to check that the outlier's score is genuine, not just an error. Check the score and see whether it is within the range of possible scores for that variable. Sometimes it is worth checking back with the questionnaire or data record to see if there was a mistake in entering the data. If it is an error, correct it, and repeat the boxplot. If it turns out to be a genuine score, you

then need to decide what you will do about it. Some statistics writers suggest removing all extreme outliers from the data file. Others take a less extreme view and suggest changing the value to a less extreme value, thus including the person in the analysis but not allowing the score to distort the statistics (for more advice on this, see Chapter 4 in Tabachnick & Fidell, 2001).

- The information in the **Descriptives** table can give you an indication of how much of a problem these outlying cases are likely to be. The value you are interested in is the 5% Trimmed Mean. To obtain this value, SPSS removed the top and bottom 5 per cent of your cases and recalculated a new mean value. If you compare the original mean and this new trimmed mean you can see if some of your more extreme scores are having a lot of influence on the mean. If these two mean values are very different, you may need to investigate these data points further. In this example, the two mean values (26.73 and 26.64) are very similar. Given this, and the fact that the values are not too different to the remaining distribution, I will retain these cases in the data file.
- If I had decided to change or remove these values, I would have inspected the **Extreme values** table. This table gives the highest and lowest values recorded for the variable and also provides the ID number of the person or case with that score. This helps to identify the case that has the outlying values. After identifying the case it would then be necessary to go back to the data file, find the particular case involved and change the outlying values (see Chapter 5).

Additional exercises

Business

Data file: *staffsurvey.sav*. See Appendix for details of the data file.

1. Follow the procedures covered in this chapter to generate *appropriate* descriptive statistics to answer the following questions:
   (a) What percentage of the staff in this organisation are permanent employees? (Use the variable *employstatus*.)
   (b) What is the average length of service for staff in the organisation? (Use the variable *service*.)
   (c) What percentage of respondents would recommend the organisation to others as a good place to work? (Use the variable *recommend*.)
2. Assess the distribution of scores on the Total Staff Satisfaction scale (*totsatis*) for employees who are permanent versus casual (*employstatus*).
   (a) Are there any outliers on this scale that you would be concerned about?
   (b) Are scores normally distributed for each group?

Health

Data file: *sleep.sav.* See Appendix for details of the data file.

1. Follow the procedures covered in this chapter to generate *appropriate* descriptive statistics to answer the following questions:
   (a) What percentage of respondents are female (*gender*)?
   (b) What is the average age of the sample?
   (c) What percentage of the sample indicated that they had a problem with their sleep (*problem*)?
   (d) What is the median number of hours sleep per weeknight (*hourwnit*)?

2. Assess the distribution of scores on the Sleepiness and Associated Sensations Scale (*totSAS*) for people who feel that they do/don't have a sleep problem (*problem*).
   (a) Are there any outliers on this scale that you would be concerned about?
   (b) Are scores normally distributed for each group?

References

Gravetter, F. J., & Wallnau, L. B. (2000). *Statistics for the behavioral sciences* (5th edn). Belmont, CA: Wadsworth.

Tabachnick, B. G., & Fidell, L. S. (2001). *Using multivariate statistics* (4th edn). New York: HarperCollins.

7 Using graphs to describe and explore the data

While the numerical values obtained in Chapter 6 provide useful information concerning your sample and your variables, some aspects are better explored visually. SPSS for Windows provides a number of different types of graphs (referred to by SPSS as charts). In this chapter I'll cover the basic procedures to obtain the following graphs:

- histograms;
- bar graphs;
- scatterplots;
- boxplots; and
- line graphs.

Spend some time experimenting with each of the different graphs and exploring their possibilities. In this chapter only a brief overview is given to get you started. To illustrate the various graphs I have used the survey.sav data file, which is included on the website accompanying this book (see p. xi and the Appendix for details). If you wish to follow along with the procedures described in this chapter you will need to start SPSS and open the file labelled survey.sav. This file can be opened only in SPSS.

At the end of this chapter instructions are also given on how to edit a graph to better suit your needs. This may be useful if you intend to use the graph in your research paper. SPSS graphs can be imported directly into your Word document. The procedure for doing this is detailed at the end of this chapter.

Histograms

Histograms are used to display the distribution of a single continuous variable (e.g. age, perceived stress scores).

Procedure for creating a histogram

1. From the menu at the top of the screen click on: **Graphs**, then click on **Histogram**.

2. Click on your variable of interest and move it into the **Variable** box. This should be a continuous variable (e.g. total perceived stress).

3. Click on **Display normal curve**. This option will give you the distribution of your variable and, superimposed over the top, how a normal curve for this distribution would look.

4. If you wish to give your graph a title click on the **Titles** button and type the desired title in the box (e.g. Histogram of Perceived Stress scores).

5. Click on **Continue**, and then **OK**.

The output generated from this procedure is shown below.

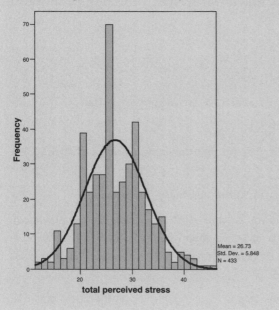

Interpretation of output from Histogram

Inspection of the shape of the histogram provides information about the distribution of scores on the continuous variable. Many of the statistics discussed in this manual assume that the scores on each of the variables are normally distributed (i.e. follow the shape of the normal curve). In this example, the scores are reasonably normally distributed, with most scores occurring in the centre, tapering out towards the extremes. It is quite common in the social sciences, however, to find that variables are not normally distributed. Scores may be skewed to the left or right or, alternatively, arranged in a rectangular shape. For further discussion of the assessment of the normality of variables, see Chapter 6.

Bar graphs

Bar graphs can be simple or very complex, depending on how many variables you wish to include. The bar graph can show the number of cases in particular categories, or it can show the score on some continuous variable for different categories. Basically you need two main variables—one categorical and one continuous. You can also break this down further with another categorical variable if you wish.

Procedure for creating a bar graph

1. From the menu at the top of the screen click on: **Graphs**, then **Bar**.

2. Click on **Clustered**.

3. In the **Data in chart are** section, click on **Summaries for groups of cases**. Click on **Define**.

4. In the **Bars represent** box, click on **Other summary function**.

5. Click on the continuous variable you are interested in (e.g. total perceived stress). This should appear in the box listed as **Mean** (Total Perceived Stress). This indicates that the mean on the Perceived Stress Scale for the different groups will be displayed.

6. Click on your first categorical variable (e.g. agegp3). Click on the arrow button to move it into the **Category axis** box. This variable will appear across the bottom of your bar graph (X axis).

7. Click on another categorical variable (e.g. sex) and move it into the **Define Clusters by:** box. This variable will be represented in the legend.

8. Click on **OK**.

The output generated from this procedure, after it has been slightly modified, is shown below.

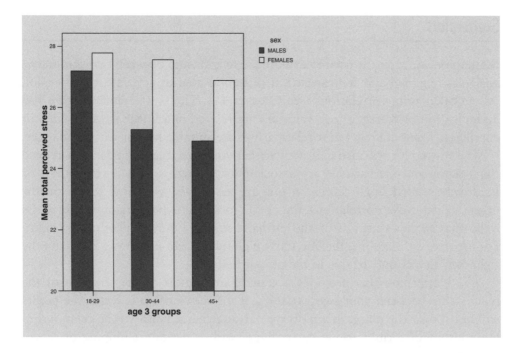

Interpretation of output from Bar Graph

The output from this procedure gives you a quick summary of the distribution of scores for the groups that you have requested (in this case, males and females from the different age groups). The graph presented above suggests that females had higher perceived stress scores than males, and that this difference is more pronounced among the two older age groups. Among the 18 to 29 age group the difference in scores between males and females is very small.

Care should be taken when interpreting the output from Bar Graph. You should always look at the scale used on the Y (vertical) axis. Sometimes what looks like a dramatic difference is really only a few scale points and, therefore, probably of little importance. This is clearly evident in the bar graph displayed above. You will see that the difference between the groups is quite small when you consider the scale used to display the graph. The difference between the smallest score (males aged 45 or more) and the highest score (females aged 18 to 29) is only about three points.

To assess the significance of any difference you might find between groups it is necessary to conduct further statistical analyses. In this case, a two-way, between-groups analysis of variance (see Chapter 18) would be conducted to find out if the differences are statistically significant.

Scatterplots

Scatterplots are typically used to explore the relationship between two continuous variables (e.g. age and self-esteem). It is a good idea to generate a scatterplot, *before* calculating correlations (see Chapter 11). The scatterplot will give you an indication of whether your variables are related in a linear (straight-line) or curvilinear fashion. Only linear relationships are suitable for correlation analyses.

The scatterplot will also indicate whether your variables are positively related (high scores on one variable are associated with high scores on the other) or negatively related (high scores on one are associated with low scores on the other). For positive correlations, the points form a line pointing upwards to the right (that is, they start low on the left-hand side and move higher on the right). For negative correlations, the line starts high on the left and moves down on the right (see an example of this in the output below).

The scatterplot also provides a general indication of the strength of the relationship between your two variables. If the relationship is weak, the points will be all over the place, in a blob-type arrangement. For a strong relationship the points will form a vague cigar shape, with a definite clumping of scores around an imaginary straight line.

In the example that follows I request a scatterplot of scores on two of the scales in the survey: the Total Perceived Stress and the Total Perceived Control of Internal States Scale (PCOISS). I have asked for two groups in my sample (males and females) to be represented separately on the one scatterplot (using different symbols). This not only provides me with information concerning my sample as a whole but also gives additional information on the distribution of scores for males and females. If you wish to obtain a scatterplot for the full sample (not split by group), just ignore the instructions below in the section labelled '**Set Markers by**'.

Procedure for creating a scatterplot

1. From the menu at the top of the screen click on: **Graphs**, then on **Scatter**.

2. Click on **Simple** and then **Define**.

3. Click on your first variable, usually the one you consider is the dependent variable, (e.g. total perceived stress).

4. Click on the arrow to move it into the box labelled **Y axis**. This variable will appear on the vertical axis.

5. Move your other variable (e.g. total PCOISS) into the box labelled **X axis**. This variable will appear on the horizontal axis.

6. You can also have SPSS mark each of the points according to some other categorical variable (e.g. sex). Move this variable into the **Set Markers by:** box. This will display males and females using different markers.

7. Move the ID variable in the **Label Cases by:** box. This will allow you to find out the ID number of a case from the graph if you find an outlier.

8. If you wish to attach a title to the graph, click on the **Titles** button. Type in the desired title and click on **Continue**.

9. Click on **OK**.

The output generated from this procedure, modified slightly for display purposes, is shown below.

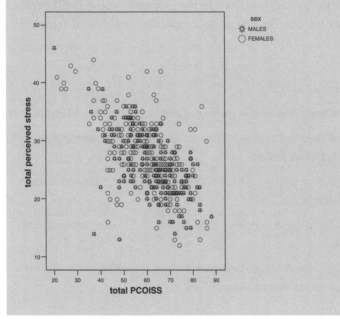

Interpretation of output from Scatterplot

From the output above, there appears to be a moderate, negative correlation between the two variables (Perceived Stress and PCOISS) for the sample as a whole. Respondents with high levels of perceived control (shown on the X, or horizontal, axis) experience lower levels of perceived stress (shown on the Y, or vertical, axis). On the other hand, people with low levels of perceived control have much greater perceived stress. There is no indication of a curvilinear relationship, so it would be appropriate to calculate a Pearson product-moment correlation for these two variables (see Chapter 11).

Remember, the scatterplot does not give you definitive answers; you need to follow it up with the calculation of the appropriate statistic (in this case, Pearson product-moment correlation coefficient).

Boxplots

Boxplots are useful when you wish to compare the distribution of scores on variables. You can use them to explore the distribution of one continuous variable (e.g. positive affect) or alternatively you can ask for scores to be broken down for different groups (e.g. age groups). You can also add an extra categorical variable to compare (e.g. males and females). In the example below I will explore the distribution of scores on the Positive Affect scale for males and females.

Procedure for creating a boxplot

1. From the menu at the top of the screen click on: **Graphs**, then click on **Boxplot**.

2. Click on **Simple**. In the **Data in Chart Are** section click on **Summaries for groups of cases**. Click on the **Define** button.

3. Click on your continuous variable (e.g. total positive affect). Click the arrow button to move it into the **Variable** box.

4. Click on your categorical variable (e.g. sex). Click on the arrow button to move into the **Category axis** box.

5. Click on ID and move it into the **Label cases** box. This will allow you to identify the ID numbers of any cases with extreme values.

6. Click on **OK**.

The output generated from this procedure is shown below.

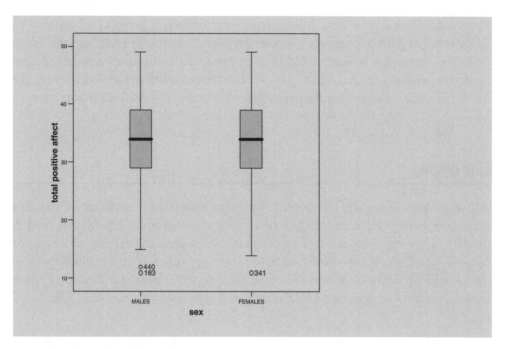

Interpretation of output from Boxplot

The output from Boxplot gives you a lot of information about the distribution of your continuous variable and the possible influence of your other categorical variable (and cluster variable if used).

- Each distribution of scores is represented by a box and protruding lines (called whiskers). The length of the box is the variable's interquartile range and contains 50 per cent of cases. The line across the inside of the box represents the median value. The whiskers protruding from the box go out to the variable's smallest and largest values.
- Any scores that SPSS considers are outliers appear as little circles with a number attached (this is the ID number of the case). Outliers are cases with scores that are quite different from the remainder of the sample, either much higher or much lower. SPSS defines points as outliers if they extend more than 1.5 box-lengths from the edge of the box. Extreme points (indicated with an asterisk, *) are those that extend more than 3 box-lengths from the edge of the box. For more information on outliers, see Chapter 6. In the example above there are a number of outliers at the low values for Positive Affect for both males and females.

- In addition to providing information on outliers, a boxplot allow you to inspect the pattern of scores for your various groups. It provides an indication of the variability in scores within each group and allows a visual inspection of the differences between groups. In the example presented above, the distribution of scores on Positive Affect for males and females is very similar.

Line graphs

A line graph allows you to inspect the mean scores of a continuous variable across a number of different values of a categorical variable (e.g. time 1, time 2, time 3). They are also useful for graphically exploring the results of a one- or two-way analysis of variance. Line graphs are provided as an optional extra in the output of analysis of variance (see Chapters 17 and 18). This procedure shows you how to generate a line graph without having to run ANOVA.

Procedure for creating a line graph

1. From the menu at the top of the screen click on: **Graphs**, then click on **Line**.

2. Click on **Multiple**. In the **Data in Chart Are** section, click on **Summaries for groups of cases**. Click on **Define**.

3. In the **Lines represent** box, click on **Other summary function**. Click on the continuous variable you are interested in (e.g. total perceived stress). Click on the arrow button. The variable should appear in the box listed as Mean (Total Perceived Stress). This indicates that the mean on the Perceived Stress Scale for the different groups will be displayed.

4. Click on your first categorical variable (e.g. agegp3). Click on the arrow button to move it into the **Category Axis** box. This variable will appear across the bottom of your line graph (X axis).

5. Click on another categorical variable (e.g. sex) and move it into the **Define Lines by**: box. This variable will be represented in the legend.

6. Click on **OK**.

The output generated from this procedure, modified slightly for display purposes, is shown below.

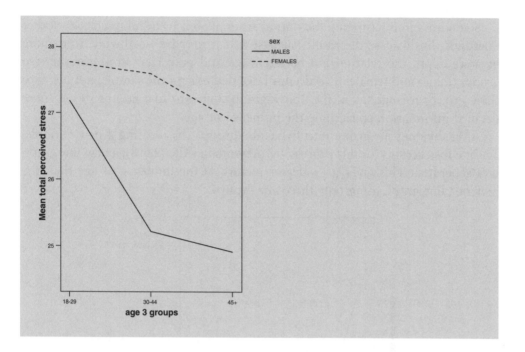

Interpretation of output from Line Graph

The line graph displayed above contains a good deal of information.

- First, you can look at the impact of age on perceived stress for each of the sexes separately. Younger males appear to have higher levels of perceived stress than either middle-aged or older males. For females the difference across the age groups is not quite so pronounced. The older females are only slightly less stressed than the younger group.
- You can also consider the difference between males and females. Overall, males appear to have lower levels of perceived stress than females. Although the difference for the younger group is only small, there appears to be a discrepancy for the older age groups. Whether or not these differences reach statistical significance can be determined only by performing a two-way analysis of variance (see Chapter 18).

The results presented above suggest that to understand the impact of age on perceived stress you must consider the respondents' gender. This sort of relationship is referred to, when doing analysis of variance, as an *interaction effect*. While the use of a line graph does not tell you whether this relationship is statistically significant, it certainly gives you a lot of information and raises a lot of additional questions.

Sometimes in interpreting the output from SPSS it is useful to consider other questions. In this case the results suggest that it may be worthwhile to explore in more depth the relationship between age and perceived stress for the two groups (males and females). To do this I decided to split the sample, not just into three groups for age, as in the above graph, but into five groups to get more detailed information concerning the influence of age.

After dividing the group into five equal groups (by creating a new variable, age5gp: instructions for this process are presented in Chapter 8), a new line graph was generated. This gives us a clearer picture of the influence of age than the previous line graph using only three age groups.

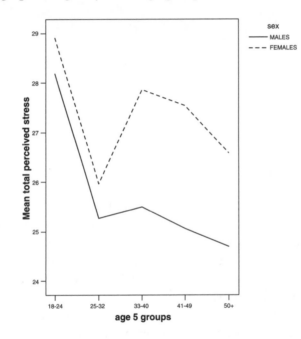

Editing a chart/graph

Sometimes modifications need to be made to the titles, labels, markers etc. of a graph before you can print it or use it in your report. For example, I have edited some of the graphs displayed in this chapter to make them clearer (e.g. changing the patterns in the bar graph, thickening the lines used in the line graph). To edit a chart or graph you need to open the **Chart Editor** window. To do this, place your cursor on the graph that you wish to modify. Double-click and a new window will appear, complete with additional menu options and icons (see Figure 7.1).

| Chart Editor |
| File Edit View Chart Help |

Figure 7.1

Example of Chart
Editor menu bar

There are a number of changes you can make while in **Chart Editor**:

- To change the words used in labels or title, click once on the title to highlight it (a blue box should appear around the text), click once again to edit the text (a red cursor should appear). Modify the text and then press Enter on your keyboard when you have finished.
- To change the position of the chart title or the X and Y axis labels (e.g. to centre them), *double-click* on the title you wish to change—in the Properties box that appears click on the Text tab. In the section labelled **Justify,** choose the position you want (the dot means centred, the left arrow moves it to the left, and the right arrow moves it to the right).
- To change the characteristics of the text, lines, markers, colours and patterns used in the chart, click *once* on the aspect of the graph that you wish to change, and then right click on your mouse and choose the **Properties** box. The various tabs in this box will allow you to change aspects of the graph. In the case where you want to change one of the lines of a multiple-line graph you will need to highlight the specific category in the legend (rather than on the graph itself). This is useful for changing one of the lines to dashes so that it is more clearly distinguishable when printed out in black and white.

The best way to learn how to use these options is to experiment—so go ahead and play!

Importing charts/graphs into Word documents

SPSS allows you to copy charts directly into your word processor (e.g. Word for Windows). This is useful when you are preparing the final version of your report and want to present some of your results in the form of a graph. Sometimes a graph will present your results more simply and clearly than numbers in a box. Don't go overboard—use only for special effect. Make sure you modify the graph to make it as clear as possible before transferring it to Word.

Please note: The instructions given below are for versions of Word running under Windows 95 or later.

Procedure for importing a chart into a Word document

Windows allows you to have more than one program open at a time. To transfer between SPSS and Word you will need to have both of these programs open. It is possible to swap backwards and forwards between the two just by clicking on the appropriate icon at the bottom of your screen. This is like shuffling pieces of paper around on your desk.

1. Start Word and open the file in which you would like the graph to appear. Click on the SPSS icon on the bottom of your screen to return to SPSS.

2. Make sure you have the Viewer window on the screen in front of you.

3. Click once on the graph that you would like to copy. A border should appear around the graph.

4. Click on **Edit** (from the menu at the top of the page) and then choose **Copy Objects**. This saves the chart to the clipboard (you won't be able to see it, however).

5. From the list of minimised programs at the bottom of your screen, click on your word processor (e.g. Microsoft Word). This will activate Word again (i.e. bring it back to the screen in front of you).

6. In the Word document place your cursor where you wish to insert the chart.

7. Click on **Edit** from the Word menu and choose **Paste**. Or just click on the **Paste** icon on the top menu bar (it looks like a clipboard).

8. Click on **File** and then **Save** to save your Word document.

9. To move back to SPSS to continue with your analyses, just click on the SPSS icon, which should be listed at the bottom of your screen. With both programs open you can just jump backwards and forwards between the two programs, copying charts, tables etc. There is no need to close either of the programs until you have finished completely. Just remember to save as you go along.

Additional exercises

Business

Data file: *staffsurvey.sav*. See Appendix for details of the data file.

1. Generate a **histogram** to explore the distribution of scores on the Staff Satisfaction Scale (*totsatis*).

2. Generate a **bar graph** to assess the staff satisfaction levels for permanent versus casual staff employed for less than or equal to 2 years, 3 to 5 years and 6 or more years. The variables you will need are *totsatis*, *employstatus* and *servicegp3*.

3. Generate a **scatterplot** to explore the relationship between years of service and staff satisfaction. Try first using the *service* variable (which is very skewed) and then try again with the variable towards the bottom of the list of variables (*logservice*). This new variable is a mathematical transformation (log 10) of the original variable (*service*), designed to adjust for the severe skewness. This procedure is covered in Chapter 8.

4. Generate a **boxplot** to explore the distribution of scores on the staff satisfaction scale (*totsatis*) for the different age groups (*age*).

5. Generate a **line graph** to compare staff satisfaction for the different age groups (use the *agerecode* variable) for permanent and casual staff.

Health

Data file: *sleep.sav*. See Appendix for details of the data file.

1. Generate a **histogram** to explore the distribution of scores on the Epworth Sleepiness Scale (*ess*).

2. Generate a **bar graph** to compare scores on the Sleepiness and Associated Sensations Scale (*totSAS*) across three age groups (*agegp3*) for males and females (*gender*).

3. Generate a **scatterplot** to explore the relationship between scores on the Epworth Sleepiness Scale (*ess*) and the Sleepiness and Associated Sensations Scale (*totSAS*). Ask for different markers for males and females (*gender*).

4. Generate a **boxplot** to explore the distribution of scores on the Sleepiness and Associated Sensations Scale (*totSAS*) for people who report that the do/don't have a problem with their sleep (*problem*).

5. Generate a **line graph** to compare scores on the Sleepiness and Associated Sensations Scale (*totSAS*) across the different age groups (use the *agegp3* variable) for males and females (*gender*).

8 Manipulating the data

Once you have entered the data and the data file has been checked for accuracy, the next step involves manipulating the raw data into a form that you can use to conduct analyses and to test your hypotheses. Depending on the data file, your variables of interest and the type of research questions that you wish to address, this process may include:

- adding up the scores from the items that make up each scale to give an overall score for scales such as self-esteem, optimism, perceived stress etc.; SPSS does this quickly, easily and accurately—don't even think about doing this by hand for each separate subject;
- transforming skewed variables for analyses that require normally distributed scores;
- collapsing continuous variables (e.g. age) into categorical variables (e.g. young, middle-aged and old) to do some analyses such as analysis of variance; and
- reducing or collapsing the number of categories of a categorical variable (e.g. collapsing the marital status into just two categories representing people 'in a relationship'/'not in a relationship').

The procedures used to manipulate the data set are described in the sections to follow.

Calculating total scale scores

Before you can perform statistical analyses on your data set you need to calculate total scale scores for any scales used in your study. This involves two steps:

- *Step 1:* reversing any negatively worded items; and
- *Step 2:* instructing SPSS to add together scores from all the items that make up the subscale or scale.

Before you commence this process it is important that you understand the scales and measures that you are using for your research. You should check with the scale's manual or the journal article it was published in to find out which items, if any, need to be reversed and how to go about calculating a total score. Some scales consist of a number of subscales which either can, or should not, be added together to give an overall score. It is important that you do this correctly, and it is much easier to do it right the first time than to have to repeat analyses later.

Step 1: Reversing negatively worded items

In some scales the wording of particular items has been reversed to help prevent response bias. This is evident in the Optimism scale used in the survey. Item 1 is worded in a positive direction (high scores indicate *high* optimism): 'In uncertain times I usually expect the best.' Item 2, however, is negatively worded (high scores indicate *low* optimism): 'If something can go wrong for me it will.' Items 4 and 6 are also negatively worded. The negatively worded items need to be reversed before a total score can be calculated for this scale. We need to ensure that all items are scored so that high scores indicate high levels of optimism. The procedure for reversing items 2, 4 and 6 of the Optimism scale is shown in the table that follows. A five-point Likert-type scale was used for the Optimism scale; therefore, scores for each item can range from 1 (strongly disagree) to 5 (strongly agree).

Warning
Make sure you have a second copy of your data set. If you make a mistake here you will lose or corrupt your original data. Therefore, it is essential that you have a backup copy. Remember that, unlike other spreadsheet programs (e.g. Excel), SPSS does not automatically recalculate values if you add extra cases or if you make changes to any of the values in the data file. Therefore you should perform the procedures illustrated in this chapter only when you have a complete (and clean) data file.

Procedure for reversing the scores of scale items

1. From the menu at the top of the screen click on: **Transform**, then click on **Recode**, then **Into Same Variables**.

2. Select the items you want to reverse (op2, op4, op6). Move these into the **Variables** box.

3. Click on the **Old and new values** button.

   In the **Old value** section, type 1 in the **Value** box.

   In the **New value** section, type 5 in the **Value** box (this will change all scores that were originally scored as 1 to a 5).

4. Click on **Add**. This will place the instruction (1—5) in the box labelled Old > New.

5. Repeat the same procedure for the remaining scores, for example:
   Old value—type in 2 **New value**—type in 4 **Add**
   Old value—type in 3 **New value**—type in 3 **Add**
   Old value—type in 4 **New value**—type in 2 **Add**
   Old value—type in 5 **New value**—type in 1 **Add**

   Always double-check the item numbers that you specify for recoding and the old and new values that you enter. Not all scales use a 5-point scale: some have 4 possible responses, some 6 and some 7. Check that you have reversed all the possible values.

6. When you are *absolutely* sure, click on **Continue** and then **OK**.

Warning. Once you click on **OK** your original data will be changed forever. In the current example all the scores on items 2, 4 and 6 of the Optimism scale, for all your subjects, will be changed permanently.

When you reverse items it is important that you make note of this in your codebook. It is also worth checking your data set to see what effect the recode had on the values. For the first few cases in your data set, take note of their scores before recoding and then check again after to ensure that it worked properly.

Step 2: Adding up the total scores for the scale

After you have reversed the negatively worded items in the scale you will be ready to calculate total scores for each subject. You should do this only when you have a complete data file.

Procedure for calculating total scale scores

1. From the menu at the top of the screen click on: **Transform**, then click on **Compute**.

2. In the **Target variable** box type in the new name you wish to give to the total scale scores (it is useful to use a T prefix to indicate total scores as this makes them easier to find in the alphabetical list of variables when you are doing your analyses).

   Important. Make sure you do not accidentally use a variable name that has already been used in the data set. If you do, you will lose all the original data— potential disaster; so check your codebook.

3. Click on the **Type and Label** button. Click in the **Label** box and type in a description of the scale (e.g. total optimism). Click on **Continue**.

4. From the list of variables on the left-hand side, click on the first item in the scale (op1).

5. Click on the arrow button > to move it into the **Numeric Expression** box.

6. Click on + on the calculator.

7. Click on the second item in the scale (op2). Click on the arrow button > to move the item into the box.

8. Click on + on the calculator and repeat the process until all scale items appear in the box.

9. The complete numeric expression should read as follows:
   op1+op2+op3+op4+op5+op6

10. Double-check that all items are correct and that there are + signs in the correct places. Click **OK**.

This will create a new variable at the end of your data set called TOPTIM. Scores for each person will consist of the addition of scores on each of the items op1 to op6. If any items had missing data, the overall score will also be missing. This is indicated by a full stop instead of a score in the data file.

You will notice in the literature that some researchers go a step further and divide the total scale score by the number of items in the scale. This can make it a little easier to interpret the scores of the total scale because it is back in the original scale used for each of the items (e.g. from 1 to 5 representing strongly disagree to strongly agree). To do this you also use the **Transform, Compute** menu of SPSS. This time you will need to specify a new variable name and then type in a suitable formula (e.g. TOPTIM/6).

Always record details of any new variables that you create in your codebook. Specify the new variable's name, what it represents and full details of what was done to calculate it. If any items were reversed, this should be specified along with details of which items were added to create the score. It is also a good idea to include the possible range of scores for the new variable in the codebook (see the Appendix). This gives you a clear guide when checking for any out-of-range values. It also helps you get a feel for the distribution of scores on your new variable. Does your mean fall in the middle of possible scores or up at one end?

After creating a new variable it is important to run **Descriptives** on this new scale to check that the values are appropriate (see Chapter 5):

- Check back with the questionnaire—what is the possible range of scores that could be recorded? For a 10-item scale, using a response scale from 1 to 4, the minimum value would be 10 and the maximum value would be 40. If a person answered 1 to every item, that overall score would be $10 \times 1 = 10$. If a person answered 4 to each item, that score would be $10 \times 4 = 40$.
- Check the output from **Descriptives** to ensure that there are no 'out-of-range' cases (see Chapter 5).
- Compare the mean score on the scale with values reported in the literature. Is your value similar to that obtained in previous studies? If not, why not? Have you done something wrong in the recoding? Or is your sample different from that used in other studies?

You should also run other analyses to check the distribution of scores on your new total scale variable:

- Check the distribution of scores using skewness and kurtosis (see Chapter 6).
- Obtain a histogram of the scores and inspect the spread of scores—are they normally distributed? If not, you may need to consider 'transforming' the scores for some analyses (this is discussed below).

Transforming variables

Often when you check the distribution of scores on a scale or measure (e.g. self-esteem, anxiety) you will find (to your dismay!) that the scores do not fall in a nice, normally distributed curve. Sometimes scores will be positively skewed, where most of the respondents record low scores on the scale (e.g. depression). Sometimes you will find a negatively skewed distribution, where most scores are at the high end (e.g. self-esteem). Given that many of the parametric statistical tests assume normally distributed scores, what do you do about these skewed distributions?

One of the choices you have is to abandon the use of parametric statistics (e.g. Pearson correlation, Analysis of Variance) and to use non-parametric alternatives (Spearman's rho, Kruskal-Wallis). SPSS includes a number of useful non-parametric techniques in its package. These are discussed in Chapter 22. Unfortunately, non-parametric techniques tend to be less 'powerful'. This means that they may not detect differences or relationships even when they actually exist.

Another alternative, when you have a non-normal distribution, is to 'transform' your variables. This involves mathematically modifying the scores using various formulas until the distribution looks more normal. There are a number of different types of transformations, depending on the shape of your distribution. There is considerable controversy concerning this approach in the literature, with some authors strongly supporting, and others arguing against, transforming variables to better meet the assumptions of the various parametric techniques. For a discussion of the issues and the approaches to transformation, you should read Chapter 4 in Tabachnick and Fidell (2001).

In Figure 8.1 some of the more common problems are represented, along with the type of transformation recommended by Tabachnick and Fidell (2001, p. 83). You should compare your distribution with those shown, and decide which picture it most closely resembles. I have also given a nasty-looking formula beside each of the suggested transformations. Don't let this throw you—these are just formulas that SPSS will use on your data, giving you a new, hopefully normally distributed variable to use in your analyses. In the procedures section to follow you will be shown how to ask SPSS to do this for you. Before attempting any of these transformations, however, it is important that you read Tabachnick and Fidell (2001, Chapter 4), or a similar text, thoroughly.

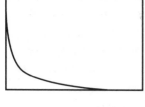

Square root
Formula: new variable = SQRT (old variable)

Figure 8.1

Distribution of scores
and suggested
transformations
(Tabachnik & Fidell,
1996)

Logarithm
Formula: new variable = LG10 (old variable)

Inverse
Formula: new variable = 1 / old variable

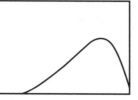

Reflect and square root
Formula: new variable = SQRT (K – old variable) where
K = largest possible value + 1

Reflect and logarithm
Formula: new variable = LG10 (K – old variable) where
K = largest possible value + 1

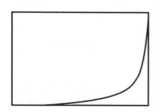

Reflect and inverse
Formula: new variable = 1 / (K – old variable) where
K = largest possible value + 1

Procedure for transforming variables

1. From the menu at the top of the screen click on: **Transform**, then click on **Compute**.

2. Target variable. In this box type in a new name for the variable. Try to include an indication of the type of transformation and the original name of the variable: e.g. for a variable called tslfest, I would make this new variable sqslfest, if I had performed a square root. Be consistent in the abbreviations that you use for each of your transformations.

3. **Functions.** Listed are a wide range of possible actions you can use. You need to choose the most appropriate transformation for your variable. Look at the shape of your distribution, compare it with those in Figure 8.1. Take note of the formula listed next to the picture that matches your distribution. This is the one that you will use.

4. **For transformations involving square root or Logarithm.** In the **Functions** box, scan down the list until you find the formula you need (e.g. SQRT or LG10). Highlight the one you want and click on the up arrow. This moves the formula into the **Numeric Expression** box. You will need to tell it which variable you want to recalculate. Find it in the list of variables and click on the arrow to move it into the expression box. If you prefer you can just type the formula in yourself without using the Functions or Variables list. Just make sure you spell everything correctly.

5. **For transformations involving Reflect.** You need to find the value K for your variable. This is the largest value that your variable can have (see your codebook) + 1. Type this number in the **Numeric Expression** box. Complete the remainder of the formula, using the **Functions** box, or alternatively type it in yourself.

6. **For transformations involving Inverse.** To calculate the inverse you need to divide your scores into 1. So in the **Numeric Expression** box type in 1 then, type / and then your variable or the rest of your formula (e.g. 1/ tslfest).

7. Check the final formula in the **Numeric Expression** box. Write this down in your codebook next to the name of the new variable you created.

8. Click on the button **Type and Label**. Under **Label** type in a brief description of the new variable (or you may choose to use the actual formula you used).

9. Check in the **Target Variable** box that you have given your new variable a *new* name, not the original one. If you accidentally put the old variable name, you will lose all your original scores. So, always double-check.

10. Click on **OK**. A new variable will be created and will appear at the end of your data file.

11. Run **Summarize**, **Descriptives** to check the skewness and kurtosis values for your new variable. Have they improved?

12. Run **Graphs**, **Histogram** to inspect the distribution of scores on your new variable. Has the distribution improved? If not, you may need to consider a different type of

transformation. If none of the transformations work you may need to consider using non-parametric techniques to analyse your data.

Collapsing a continuous variable into groups

For some analyses (e.g. Analysis of Variance) you may wish to divide the sample into equal groups according to respondents' scores on some variable (e.g. to give low, medium and high scoring groups). In the previous versions of SPSS (before SPSS Version 12) this required a number of steps (identifying cut-off points, and then recoding scores into a new variable). This new version of SPSS (Version 12) has an option (**Visual Bander**) under its **Transform** menu that will do all the hard work for you.

To illustrate this process, I will use the survey.sav file that is included on the website that accompanies this book (see p. xi and the Appendix for details). I will use **Visual Bander** to identify suitable cut-off points to break the continuous variable *age* into three approximately equal groups. The same technique could be used to create a 'median split': that is, to divide the sample into two groups, using the median as the cut-off point. Once the cut-off points are identified, **Visual Bander** will create a new (additional) categorical variable that has only three values, corresponding to the three age ranges chosen. This technique leaves the original variable age, measured as a continuous variable, intact so that you can use it for other analyses.

Procedure for collapsing a continuous variable into groups

1. From the menu at the top of the screen click on: **Transform**, and choose **Visual Bander.**

2. Select the continuous variable that you want to use (e.g. age). Transfer it into the **Variables to Band** box. Click on the **Continue** button.

3. In the Visual Bander screen that appears, click on the variable to highlight it. A histogram, showing the distribution of age scores should appear.

4. At the section at the top labelled **Banded Variable** type in the name for the new categorical variable that you will create (e.g. Ageband3). You can also change the suggested label that is shown (e.g. to age in 3 groups).

5. Click on button labelled **Make Cutpoints**, and then **OK**. In the dialogue box that appears click on the option **Equal Percentiles Based on Scanned Cases**. In the box **Number of Cutpoints** specify a number one less than the number of groups that you want (e.g. if you want three groups, type in 2 for cutpoints). In the **Width (%)** section below you will then see 33.33 appear—this means that

> SPSS will try to put 33.3 per cent of the sample in each group. Click on the **Apply** button.
>
> 6. Click on the **Make Labels** button back in the main dialogue box. This will automatically generate value labels for each of the new groups created. You can modify these if you wish by clicking in the cells of the Grid.
>
> 7. Click on **OK** and a new variable (Ageband3) will appear at the end of your data file (go back to your **Data Editor** window, choose the **Variable View** tab, and it should be at the bottom).
>
> 8. Run **Descriptives, Frequencies** on your newly created variable (Ageband3) to check the number of cases in each of the categories.

Collapsing the number of categories of a categorical variable

There are some situations where you may want or need to reduce or collapse the number of categories of a categorical variable. You may want to do this for research or theoretical reasons (e.g. collapsing the marital status into just two categories representing people 'in a relationship'/'not in a relationship') or you may make the decision after looking at the data.

After running **Descriptive Statistics** you may find you have only a few people in your sample that fall into a particular category (e.g. for our education variable we only have two people in our first category, 'primary school'). As it stands this variable could not appropriately be used in many of the statistical analyses covered later in the book. We could decide just to remove these people from the sample, or we could recode them to combine them with the next category (some secondary school). We would have to relabel the variable so that it represented people who did not complete secondary school. The procedure for recoding a categorical variable is shown below. It is very important to note that here we are creating a new additional variable (so that we keep our original data intact).

Procedure for recoding a categorical variable

1. From the menu at the top of the screen click on **Transform**, then on **Recode**, then on **Into Different Variables**. (Make sure you select 'different variables', as this retains the original variable for other analyses).

2. Select the variable you wish to recode (e.g. educ). In the **Name** box type a name for the new variable that will be created (e.g. educrec). Type in an extended label if you wish in the **Label** section. Click on the button labelled **Change**.

3. Click on the button labelled **Old and New Values**.

4. In the section **Old Value** you will see a box labelled **Value**. Type in the first code or value of your current variable (e.g. 1). In the **New Value** section type in the new value that will be used (or, if the same one is to be used type that in). In this case I will recode to the same value, so I will type 1 in both the **Old Value** and **New Value** section. Click on the **Add** button.

5. For the second value I would type 2 in the **Old Value**, but in the **New Value** I would type 1. This will recode all the values of both 1 and 2 from the original coding into one group in the new variable to be created with a value of 1.

6. For the third value of the original variable I would type 1 in the **Old Value** and 2 in the **New Value**. This is just to keep the values in the new variable in sequence. Click on **Add**. Repeat for all the remaining values of the original values. In the table Old→New you should see the following codes for this example: 1→1; 2→1; 3→2; 4→3; 5→4; 6→5.

7. Click on **Continue.**

8. Go to your **Data Editor** window and choose the **Variable View** tab. Type in appropriate values labels to represent the new values (1=did not complete high school, 2=completed high school, 3=some additional training, 4=completed undergrad uni, 5=completed postgrad uni). Remember, these will be different from the codes used for the original variable, and it is important that you don't mix them up.

When you recode a variable, make sure you run **Frequencies** on both the old variable (educ) and the newly created variable (educrec: which appears at the end of your data file). Check that the frequencies reported for the new variable are correct. For example, for the newly created Educrec variable we should now have 2+53=55 in the first group. This represents the 2 people who ticked 1 on the original variable (primary school), also the 53 people who ticked 2 (some secondary school).

The Recode procedure demonstrated here could be used for a variety of purposes. You may find later when you come to do your statistical analyses you will need to recode the values used for a variable. For example, in the Logistic Regression chapter (Chapter 14) you may need to recode variables originally coded 1=yes, 2=no to a new coding system 1=yes, 0=no. This can be achieved in the same way as described in the previous procedures section. Just be very clear before you start about the original values, and about what you want the new values to be.

Additional exercises

Business

Data file: *staffsurvey.sav*. See Appendix for details of the data file.

1. Practise the procedures described in this chapter to add up the total scores for a scale using the items that make up the Sstaff Satisfaction Survey. You will need to add together the items that assess agreement with each item in the scale (i.e. Q1a+Q2a+Q3a . . . to Q10a). Name your new variable *staffsatis*.

2. Check the descriptive statistics for your new total score (*staffsatis*) and compare this with the descriptives for the variable *totsatis* which is already in your datafile. This is the total score that I have already calculated for you.

3. What is the minimum possible and maximum possible scores for this new variable? Tip: check the number of items in the scale and the number of response points on each item (see Appendix).

4. Check the distribution of the variable *service* by generating a histogram. You will see that it is very skewed, with most people clustered down the low end (with less than 2 years service), and a few people stretched up at the very high end (with more than 30 years service). Check the shape of the distribution against those displayed in Figure 8.1 and try a few different transformations. Remember to check the distribution of the new transformed variables you create. Are any of the new variables more 'normally' distributed?

5. Collapse the years of service variable (*service*) into three groups using the **Visual Bander** procedure from the **Transform** menu. Use the **Make Cutpoints** button and ask for **Equal Percentiles**. In the section labelled **Number of Cutpoints** specify 2. Call your new variable *gp3service* to distinguish it from the variable I have already created in the datafile using this procedure (*service3gp*). Run **Frequencies** on your newly created variable to check how many cases are in each group.

Health

Data file: *sleep.sav*. See Appendix for details of the data file.

1. Practise the procedures described in this chapter to add up the total scores for a scale using the items that make up the Sleepiness and Associated Sensations Scale. You will need to add together the items *fatigue, lethargy, tired, sleepy, energy*. Call your new variable *sleeptot*. Please note: none of these items needs to be reversed before being added.

2. Check the descriptive statistics for your new total score (*sleeptot)* and compare this with the descriptives for the variable *totSAS* which is already in your datafile. This is the total score that I have already calculated for you.

3. What is the minimum possible and maximum possible scores for this new variable. Tip: check the number of items in the scale and the number of response points on each item (see Appendix).

4. Check the distribution (using a histogram) of the variable which measures the number of cigarettes smoked per day by the smokers in the sample (*smokenum*). You will see that it is very skewed, with most people clustered down the low end (with less than 10 per day) and a few people stretched up at the very high end (with more than 70 per day). Check the shape of the distribution against those displayed in Figure 8.1 and try a few different transformations. Remember to check the distribution of the new transformed variables you create. Are any of the new transformed variables more 'normally' distributed?

5. Collapse the age variable (*age*) into three groups using the **Visual Bander** procedure from the **Transform** menu. Use the **Make Cutpoints** button and ask for **Equal Percentiles**. In the section labelled **Number of Cutpoints** specify 2. Call your new variable *gp3age* to distinguish it from the variable I have already created in the datafile using this procedure (*age3gp*). Run **Frequencies** on your newly created variable to check how many cases are in each group.

Reference

Tabachnick, B. G., & Fidell, L. S. (2001). *Using multivariate statistics* (4th edn). New York: HarperCollins.

9 Checking the reliability of a scale

When you are selecting scales to include in your study it is important to find scales that are reliable. There are a number of different aspects to reliability (see discussion of this in Chapter 1). One of the main issues concerns the scale's internal consistency. This refers to the degree to which the items that make up the scale 'hang together'. Are they all measuring the same underlying construct? One of the most commonly used indicators of internal consistency is Cronbach's alpha coefficient. Ideally, the Cronbach alpha coefficient of a scale should be above .7. Cronbach alpha values are, however, quite sensitive to the number of items in the scale. With short scales (e.g. scales with fewer than ten items), it is common to find quite low Cronbach values (e.g. .5). In this case it may be more appropriate to report the mean inter-item correlation for the items. Briggs and Cheek (1986) recommend an optimal range for the inter-item correlation of .2 to .4.

The reliability of a scale can vary depending on the sample that it is used with. It is therefore necessary to check that each of your scales is reliable with your particular sample. This information is usually reported in the Method section of your research paper or thesis. If your scale contains some items that are negatively worded (common in psychological measures), these need to be 'reversed' *before* checking reliability. Instructions for how to do this are provided in Chapter 8. Before proceeding, make sure that you check with the scale's manual (or the journal article that it is reported in) for instructions concerning the need to reverse items and also for information on any subscales. Sometimes scales contain a number of subscales that may, or may not, be combined to form a total scale score. If necessary, the reliability of each of the subscales and the total scale will need to be calculated.

Details of example

To demonstrate this technique I will be using the survey.sav data file included on the website accompanying this book (see p. xi). Full details of the study, the questionnaire and scales used are provided in the Appendix. If you wish to follow along with the steps described in this chapter you should start SPSS and open the file labelled survey.sav. This file can be opened only in SPSS. In the procedure described below I will explore the internal consistency of one of the scales from the questionnaire. This is the Satisfaction with Life scale, which is made up of five items. These items are labelled in the data file as follows: lifsat1, lifsat2, lifsat3, lifsat4, lifsat5.

Procedure for checking the reliability of a scale

Important: Before starting, you should check that all negatively worded items in your scale have been reversed (see Chapter 8). If you don't do this you will find you have very low (and incorrect) Cronbach alpha values.

1. From the menu at the top of the screen click on: **Analyze**, then click on **Scale**, then **Reliability Analysis**.

2. Click on all of the individual items that make up the scale (e.g. lifsat1, lifsat2, lifsat3, lifsat4, lifsat5). Move these into the box marked **Items**.

3. In the **Model** section, make sure **Alpha** is selected.

4. Click on the **Statistics** button. In the **Descriptives for** section, click on **Item**, **Scale**, and **Scale if item deleted**.

5. Click on **Continue** and then **OK**.

The output generated from this procedure is shown below.

Case Processing Summary

		N	%
Cases	Valid	436	99.3
	Excluded[a]	3	.7
	Total	439	100.0

a. Listwise deletion based on all variables in the procedure.

Reliability Statistics

Cronbach's Alpha	N of Items
.890	5

Item Statistics

	Mean	Std. Deviation	N
lifsat1	4.37	1.528	436
lifsat2	4.57	1.554	436
lifsat3	4.69	1.519	436
lifsat4	4.75	1.641	436
lifsat5	3.99	1.855	436

Item-Total Statistics

	Scale Mean if Item Deleted	Scale Variance if Item Deleted	Corrected Item-Total Correlation	Cronbach's Alpha if Item Deleted
lifsat1	18.00	30.667	.758	.861
lifsat2	17.81	30.496	.752	.862
lifsat3	17.69	29.852	.824	.847
lifsat4	17.63	29.954	.734	.866
lifsat5	18.39	29.704	.627	.896

Scale Statistics

Mean	Variance	Std. Deviation	N of Items
22.38	45.827	6.770	5

Interpreting the output from reliability

The output provides you with a number of pieces of information concerning your scale.

- The first thing you should do is check that the number of items is correct. Also check that the mean score is what you would expect from the range of possible scores. Errors in recoding a variable can result in major problems here.
- In terms of reliability the most important figure is the **Alpha** value. This is Cronbach's alpha coefficient, which in this case is .89. This value is above .7, so the scale can be considered reliable with our sample.
- The other information of interest is the column marked **Corrected Item-Total Correlation**. These figures give you an indication of the degree to which each item correlates with the total score. Low values (less than .3) here indicate that the item is measuring something different from the scale as a whole. If your scale's overall Cronbach alpha is too low (e.g. less than .7) you may need to consider removing items with low item-total correlations. In the column headed **Alpha if Item Deleted,** the impact of removing each item from the scale is given. Compare these values with the final alpha value obtained. If any of the values in this column are higher than the final alpha value, you may want to consider removing this item from the scale. For established, well-validated scales, you would normally consider doing this only if your alpha value was low (less than .7).

Presenting the results from reliability

You would normally report the internal consistency of the scales that you are using in your research in the Method section of your report, under the heading Measures, or Materials. You should include a summary of reliability information reported by the scale developer and other researchers, and then a sentence to indicate the results for your sample. For example:

According to Pavot, Diener, Colvin and Sandvik (1991), the Satisfaction with Life scale has good internal consistency, with a Cronbach alpha coefficient reported of .85. In the current study the Cronbach alpha coefficient was .89.

Additional exercises

Business

Data file: *staffsurvey.sav*. See Appendix for details of the data file.

1. Check the reliability of the Staff Satisfaction Survey which is made up of the agreement items in the datafile: Q1a to Q10a. None of the items of this scale needs to be reversed.

Health

Data file: *sleep.sav*. See Appendix for details of the data file.

1. Check the reliability of the Sleepiness and Associated Sensations Scale which is made up of items *fatigue, lethargy, tired, sleepy, energy*. None of the items of this scale needs to be reversed.

References

For a simple, easy-to-follow summary of reliability and other issues concerning scale development and evaluation, see:

Oppenheim, A. N. (1992). *Questionnaire design, interviewing and attitude measurement.* London: Pinter.

For a more detailed coverage of the topic, see:

Briggs, S. R., & Cheek, J. M. (1986). The role of factor analysis in the development and evaluation of personality scales. *Journal of Personality, 54,* 106–148.
DeVellis, R. F. (1991). *Scale development: Theory and applications.* Newbury, CA: Sage.
Gable, R. K., & Wolf, M. B. (1993). *Instrument development in the affective domain: Measuring attitudes and values in corporate and school settings.* Boston: Kluwer Academic.
Kline, P. (1986). *A handbook of test construction.* New York: Methuen.
Streiner, D. L., & Norman, G. R. (1995). *Health measurement scales: A practical guide to their development and use* (2nd edn). Oxford: Oxford University Press.

For Satisfaction with Life scale, see:

Pavot, W., Diener, E., Colvin, C. R., & Sandvik, E. (1991). Further validation of the Satisfaction with Life scale: Evidence for the cross method convergence of well being measures. *Journal of Personality Assessment, 57,* 149–161.

10 Choosing the right statistic

One of the most difficult (and potentially fear-inducing) parts of the research process for most research students is choosing the correct statistical technique to analyse their data. Although most statistics courses teach you how to calculate a correlation coefficient or perform a t-test, they typically do not spend much time helping students learn how to choose which approach is appropriate to address particular research questions. In most research projects it is likely that you will use quite a variety of different types of statistics, depending on the question you are addressing and the nature of the data that you have. It is therefore important that you have at least a basic understanding of the different statistics, the type of questions they address and their underlying assumptions and requirements.

So, dig out your statistics texts and review the basic techniques and the principles underlying them. You should also look through journal articles on your topic and identify the statistical techniques used in these studies. Different topic areas may make use of different statistical approaches, so it is important that you find out what other researchers have done in terms of data analysis. Look for long, detailed journal articles that clearly and simply spell out the statistics that were used. Collect these together in a folder for handy reference. You might also find them useful later when considering how to present the results of your analyses.

In this chapter we will look at the various statistical techniques that are available and I will then take you step by step through the decision-making process. If the whole statistical process sends you into a panic, just think of it as choosing which recipe you will use to cook dinner tonight. What ingredients do you have in the refrigerator, what type of meal do you feel like (soup, roast, stir-fry, stew), and what steps do you have to follow? In statistical terms we will look at the type of research questions you have, which variables you want to analyse, and the nature of the data itself.

If you take this process step by step you will find the final decision is often surprisingly simple. Once you have determined what you have, and what you want to do, there often is only one choice. The most important part of this whole process is clearly spelling out what you have, and what you want to do with it.

Overview of the different statistical techniques

This section is broken into two main parts. First, we will look at the techniques used to explore the *relationship among variables* (e.g. between age and optimism), followed by techniques you can use when you want to explore the *differences between groups* (e.g. sex differences in optimism scores). I have separated the techniques into these two sections, as this is consistent with the way in which most basic statistics texts are structured and how the majority of students will have been taught basic statistics. This tends to somewhat artificially emphasise the difference between these two groups of techniques. There are, in fact, many underlying similarities between the various statistical techniques, which is perhaps not evident on initial inspection. A full discussion of this point is beyond the scope of this book. If you would like to know more, I would suggest you start by reading Chapter 17 of Tabachnick and Fidell (2001). That chapter provides an overview of the General Linear Model, under which many of the statistical techniques can be considered.

I have deliberately kept the summaries of the different techniques brief and simple, to aid initial understanding. This chapter certainly does not cover all the different techniques available, but it does give you the basics to get you started and to build your confidence.

Exploring relationships

Often in survey research you will not be interested in differences between groups, but instead in the strength of the relationship between variables. There are a number of different techniques that you can use.

Pearson correlation

Pearson correlation is used when you want to explore the strength of the relationship between two continuous variables. This gives you an indication of both the direction (positive or negative) and the strength of the relationship. A positive correlation indicates that as one variable increases, so does the other. A negative correlation indicates that as one variable increases, the other decreases. This topic is covered in Chapter 11.

Partial correlation

Partial correlation is an extension of Pearson correlation—it allows you to control for the possible effects of another confounding variable. Partial correlation 'removes' the effect of the confounding variable (e.g. socially desirable responding), allowing you to get a more accurate picture of the relationship between your two variables of interest. Partial correlation is covered in Chapter 12.

Multiple regression

Multiple regression is a more sophisticated extension of correlation and is used when you want to explore the predictive ability of a set of independent variables on one *continuous* dependent measure. Different types of multiple regression allow you to compare the predictive ability of particular independent variables and to find the best set of variables to predict a dependent variable. See Chapter 13.

Factor analysis

Factor analysis allows you to condense a large set of variables or scale items down to a smaller, more manageable number of dimensions or factors. It does this by summarising the underlying patterns of correlation and looking for 'clumps' or groups of closely related items. This technique is often used when developing scales and measures, to identify the underlying structure. See Chapter 15.

Summary

All of the analyses described above involve exploration of the relationship between continuous variables. If you have only categorical variables, you can use the chi-square test for relatedness or independence to explore their relationship (e.g. if you wanted to see whether gender influenced clients' dropout rates from a treatment program). In this situation you are interested in the number of people in each category (males and females, who drop out of/complete the program), rather than their score on a scale.

Some additional techniques you should know about, but which are not covered in this text, are described below. For more information on these, see Tabachnick and Fidell (2001). These techniques are as follows:

- *Discriminant function analysis* is used when you want to explore the predictive ability of a set of independent variables, on one *categorical* dependent measure. That is, you want to know which variables best predict group membership. The dependent variable in this case is usually some clear criterion (passed/failed, dropped out of/continued with treatment). See Chapter 11 in Tabachnick and Fidell (2001).
- *Canonical correlation* is used when you wish to analyse the relationship between two *sets* of variables. For example, a researcher might be interested in how a variety of demographic variables relate to measures of wellbeing and adjustment. See Chapter 6 in Tabachnick and Fidell (2001).
- *Structural equation modelling* is a relatively new, and quite sophisticated, technique that allows you to test various models concerning the inter-relationships among a set of variables. Based on multiple regression and factor analytic techniques, it allows you to evaluate the importance of each of the independent variables in the model and to test the overall fit of the model to your data. It also allows you to compare alternative models. SPSS does not have a structural equation modelling module, but it does support an 'add on' called AMOS. See Chapter 14 in Tabachnick and Fidell (2001).

Exploring differences between groups

There is another family of statistics that can be used when you want to find out whether there is a statistically significant difference among a number of groups.

Most of these analyses involve comparing the mean score for each group on one or more dependent variables. There are a number of different but related statistics in this group. The main techniques are very briefly described below.

T-tests

T-tests are used when you have *two* groups (e.g. males and females) or two sets of data (before and after), and you wish to compare the mean score on some continuous variable. There are two main types of t-tests. Paired sample t-tests (also called repeated measures) are used when you are interested in changes in scores for subjects tested at Time 1, and then again at Time 2 (often after some intervention or event). The samples are 'related' because they are the *same* people tested each time. Independent sample t-tests are used when you have two *different* (independent) groups of people (males and females), and you are interested in comparing their scores. In this case you collect information on only one occasion, but from two different sets of people. T-tests are covered in Chapter 16.

One-way analysis of variance

One-way analysis of variance is similar to a t-test, but is used when you have *two or more groups* and you wish to compare their mean scores on a continuous variable. It is called one-way because you are looking at the impact of only one independent variable on your dependent variable. A one-way analysis of variance (ANOVA) will let you know whether your groups differ, but it won't tell you where the significant difference is (gp1/gp3, gp2/gp3 etc.). You can conduct post-hoc comparisons to find out which groups are significantly different from one another. You could also choose to test differences between specific groups, rather than comparing all the groups, by using planned comparisons. Similar to t-tests, there are two types of one-way ANOVAs: repeated measures ANOVA (same people on more than two occasions), and between-groups (or independent samples) ANOVA, where you are comparing the mean scores of two or more different groups of people. One-way ANOVA is covered in Chapter 17.

Two-way analysis of variance

Two-way analysis of variance allows you to test the impact of two independent variables on one dependent variable. The advantage of using a two-way ANOVA is that it allows you to test for an interaction effect—that is, when the effect of one independent variable is influenced by another; for example, when you suspect that optimism increases with age, but only for males. It also tests for 'main effects'—that is, the overall effect of each independent variable (e.g. sex, age). There are two different two-way ANOVAs: between-groups ANOVA (when the groups are different) and repeated measures ANOVA (when the same people are tested on more than one occasion). Some research designs combine both between-groups and repeated measures in the one study. These are referred to as 'Mixed Between-Within Designs', or 'Split Plot'. Two-way ANOVA is covered in Chapter 18, mixed designs are covered in Chapter 19.

Multivariate analysis of variance

Multivariate analysis of variance (MANOVA) is used when you want to compare your groups on a number of different, but *related*, dependent variables: for example, comparing the effects of different treatments on a variety of outcome measures (e.g. anxiety, depression, physical symptoms). Multivariate ANOVA can be used with one-way, two-way and higher factorial designs involving one, two, or more independent variables. MANOVA is covered in Chapter 20.

Analysis of covariance

Analysis of covariance (ANCOVA) is used when you want to statistically control for the possible effects of an additional confounding variable (covariate). This is useful when you suspect that your groups differ on some variable that may influence the effect that your independent variables have on your dependent variable. To be sure that it is the independent variable that is doing the influencing, ANCOVA statistically removes the effect of the covariate. Analysis of covariance can be used as part of a one-way, two-way or multivariate design. ANCOVA is covered in Chapter 21.

The decision-making process

Having had a look at the variety of choices available, it is time to choose which techniques are suitable for your needs. In choosing the right statistic you will need to consider a number of different factors. These include consideration of the type of question you wish to address, the type of items and scales that were included in your questionnaire, the nature of the data you have available for each of your variables and the assumptions that must be met for each of the different statistical techniques. I have set out below a number of steps that you can use to navigate your way through the decision-making process.

Step 1: What questions do you want to address?

Write yourself a full list of all the questions you would like to answer from your research. You might find that some questions could be asked a number of different ways. For each of your areas of interest, see if you can present your question in a number of different ways. You will use these alternatives when considering the different statistical approaches you might use.

For example, you might be interested in the effect of age on optimism. There are a number of ways you could ask the question:

- Is there a relationship between age and level of optimism?
- Are older people more optimistic than younger people?

These two different questions require different statistical techniques. The question of which is more suitable may depend on the nature of the data you have collected. So, for each area of interest, detail a number of different questions.

Step 2: Find the questionnaire items and scales that you will use to address these questions

The type of items and scales that were included in your study will play a large part in determining which statistical techniques are suitable to address your research questions. That is why it is so important to consider the analyses that you intend to use when first designing your study.

For example, the way in which you collected information about respondents' age (see example in Step 1) will determine which statistics are available for you to use. If you asked people to tick one of two options (under 35/over 35), your choice of statistics would be very limited because there are only two possible values for your variable age. If, on the other hand, you asked people to give their age in years, your choices are broadened because you can have scores varying across a wide range of values, from 18 to 80+. In this situation you may choose to collapse the range of ages down into a smaller number of categories for some analyses (ANOVA), but the full range of scores is also available for other analyses (e.g. correlation).

If you administered a questionnaire or survey for your study, go back to the specific questionnaire items and your codebook and find each of the individual questions (e.g. age) and total scale scores (e.g. optimism) that you will use in your analyses. Identify each variable, how it was measured, how many response options there were and the possible range of scores.

If your study involved an experiment, check how each of your dependent and independent variables were measured. Did the scores on the variable consist of the number of correct responses, an observer's rating of a specific behaviour, or the length of time a subject spent on a specific activity? Whatever the nature of the study, just be clear that you know how each of your variables was measured.

Step 3: Identify the nature of each of your variables

The next step is to identify the nature of each of your variables. In particular, you need to determine whether each of your variables is (a) an independent variable or (b) a dependent variable. This information comes not from your data but from your understanding of the topic area, relevant theories and previous research. It is essential that you are clear in your own mind (and in your research questions) concerning the relationship between your variables—which ones are doing the influencing (independent) and which ones are being affected (dependent). There are some analyses (e.g. correlation) where it is not necessary to specify which variables are independent and dependent. For other analyses, such as ANOVA, it is important that you have this clear. Drawing a model of how you see your variables relating is often useful here (see Step 4 discussed next).

It is also important that you know the level of measurement for each of your variables. Different statistics are required for variables that are categorical and continuous, so it is important to know what you are working with. Are your variables:

- categorical (also referred to as nominal level data, e.g. sex: male/females)
- ordinal (rankings: 1st, 2nd, 3rd); and
- continuous (also referred to as interval level data, e.g. age in years, or scores on the Optimism scale).

There are some occasions where you might want to change the level of measurement for particular variables. You can 'collapse' continuous variable responses down into a smaller number of categories (see Chapter 8). For example, age can be broken down into different categories (e.g. under 35/over 35). This can be useful if you want to conduct an ANOVA. It can also be used if your continuous variables do not meet some of the assumptions for particular analyses (e.g. very skewed distributions). Summarising the data does have some disadvantages, however, as you lose information. By 'lumping' people together you can sometimes miss important differences. So you need to weigh up the benefits and disadvantages carefully.

Additional information required for continuous and categorical variables

For *continuous* variables you should collect information on the distribution of scores (e.g. are they normally distributed or are they badly skewed?). What is the range of scores? (See Chapter 6 for the procedures to do this.)

If your variable involves *categories* (e.g. group 1/group 2, males/females) find out how many people fall into each category (are the groups equal or very unbalanced?). Are some of the possible categories empty? (See Chapter 6.)

All of this information that you gather about your variables here will be used later to narrow down the choice of statistics to use.

Step 4: Draw a diagram for each of your research questions

I often find that students are at a loss for words when trying to explain what they are researching. Sometimes it is easier, and clearer, to summarise the key points in a diagram. The idea is to pull together some of the information you have collected in Steps 1 and 2 above in a simple format that will help you choose the correct statistical technique to use, or to choose among a number of different options.

One of the key issues you should be considering is: Am I interested in the *relationship* between two variables, or am I interested in *comparing* two groups of subjects? Summarising the information that you have, and drawing a diagram for each question, may help clarify this for you. I will demonstrate by setting out the information and drawing diagrams for a number of different research questions.

Question 1: Is there a relationship between age and level of optimism?
Variables:
- Age—continuous: age in years from 18 to 80; and
- Optimism—continuous: scores on the Optimism scale, ranging from 6 to 30.

From your literature review you hypothesise that, as age increases, so too will optimism levels. This relationship between two continuous variables could be illustrated as follows:

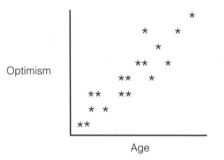

If you expected optimism scores to increase with age, you would place the points starting low on the left and moving up towards the right. If you predicted that optimism would decrease with age, then your points would start high on the left-hand side and would fall as you moved towards the right.

Question 2: Are males more optimistic than females?
Variables:
- Sex—independent, categorical (two groups): males/females; and
- Optimism—dependent, continuous: scores on the Optimism scale, range from 6 to 30.

The results from this question, with one categorical variable (with only two groups) and one continuous variable, could be summarised as follows:

	Males	Females
Mean optimism score		

Question 3: Is the effect of age on optimism different for males and females?

If you wished to investigate the joint effects of age and gender on optimism scores you might decide to break your sample into three age groups (under 30, 31–49 years and 50+).

Variables:
- Sex—independent, categorical: males/females;
- Age—independent, categorical: subjects divided into three equal groups; and
- Optimism—dependent, continuous: scores on the Optimism scale, range from 6 to 30.

The diagram might look like this:

		Age		
		Under 30	**31–49**	**50 years and over**
Mean optimism score	Males			
	Females			

Question 4: How much of the variance in life satisfaction can be explained by a set of personality factors (self-esteem, optimism, perceived control)?

Perhaps you are interested in comparing the predictive ability of a number of different independent variables on a dependent measure. You are also interested in how much variance in your dependent variable is explained by the set of independent variables.

Variables:
- Self-esteem—independent, continuous;
- Optimism—independent, continuous;
- Perceived control—independent, continuous; and
- Life satisfaction—dependent, continuous.

Your diagram might look like this:

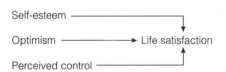

Step 5: Decide whether a parametric or a non-parametric statistical technique is appropriate

Just to confuse poor research students even further, the wide variety of statistical techniques that are available are classified into two main groups: parametric and non-parametric. Parametric statistics are more powerful, but they do have more 'strings attached': that is, they make assumptions about the data that are more stringent. For example, they assume that the underlying distribution of scores in the population from which you have drawn your sample is normal.

Each of the different parametric techniques (such as t-tests, ANOVA, Pearson correlation) has other additional assumptions. It is important that you check these *before* you conduct your analyses. The specific assumptions are listed for each of the techniques covered in the remaining chapters of this book.

What if you don't meet the assumptions for the statistical technique that you want to use? Unfortunately, in social science research, this is a common situation. Many of the attributes we want to measure are in fact not normally distributed. Some are strongly skewed, with most scores falling at the low end (e.g. depression), others are skewed so that most of the scores fall at the high end of the scale (e.g. self-esteem).

If you don't meet the assumptions of the statistic you wish to use, you have a number of choices, and these are detailed below.

Option 1

You can use the parametric technique anyway and hope that it does not seriously invalidate your findings. Some statistics writers argue that most of the approaches are fairly 'robust': that is, they will tolerate minor violations of assumptions, particularly if you have a good size sample. If you decide to go ahead with the analysis anyway, you will need to justify this in your write-up, so collect together useful quotes from statistics writers, previous researchers etc. to support your decision. Check journal articles on your topic area, particularly those that have used the same scales. Do they mention similar problems? If so, what have these other authors done? For a simple, easy-to-follow review of the robustness of different tests, see Cone and Foster (1993).

Option 2

You may be able to manipulate your data so that the assumptions of the statistical test (e.g. normal distribution) are met. Some authors suggest 'transforming' your variables if their distribution is not normal (see Chapter 8). There is some controversy concerning this approach, so make sure you read up on this so that you can justify what you have done (see Tabachnick & Fidell, 2001).

Option 3

The other alternative when you really don't meet parametric assumptions is to use a non-parametric technique instead. For many of the commonly used parametric techniques there is a corresponding non-parametric alternative. These still come with some assumptions but less stringent ones. These non-parametric alternatives (e.g. Kruskal-Wallis, Mann-Whitney U, chi-square) tend to be not as powerful: that is, they may be less sensitive in detecting a relationship, or a difference among groups. Some of the more commonly used non-parametric techniques are covered in Chapter 22.

Step 6: Making the final decision

Once you have collected the necessary information concerning your research questions, the level of measurement for each of your variables and the characteristics of the data you have available, you are finally in a position to consider your options. In the text below, I have summarised the key elements of some of the major statistical approaches you are likely to encounter. Scan down the list, find an example of the type of research question you want to address and check that you have all the necessary ingredients. Also consider whether there might be other ways you could ask your question and use a different statistical approach. I have included a summary table at the end of this chapter to help with the decision-making process.

Seek out additional information on the techniques you choose to use to ensure that you have a good understanding of their underlying principles and their assumptions. It is a good idea to use a number of different sources for this process: different authors have different opinions. You should have an understanding of the controversial issues—you may even need to justify the use of a particular statistic in your situation—so make sure you have read widely.

Key features of the major statistical techniques

This section is divided into two sections:

1. techniques used to explore relationships among variables (covered in Part Four of this book); and
2. techniques used to explore differences among groups (covered in Part Five of this book).

Exploring relationships among variables

Chi-square for independence

Example of research question:	What is the relationship between gender and dropout rates from therapy?
What you need:	• one categorical independent variable (e.g. sex: males/females); and
	• one categorical dependent variable (e.g. dropout: Yes/No).
	You are interested in the *number* of people in each category (not scores on a scale).

Diagram:

		Males	**Females**
Dropout	Yes		
	No		

Correlation

Example of research question: Is there a relationship between age and optimism scores? Does optimism increase with age?

What you need: two continuous variables (e.g., age, optimism scores)

Diagram:

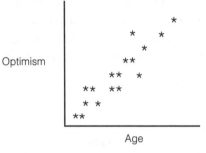

Non-parametric alternative: Spearman's Rank Order Correlation

Partial correlation

Example of research question: After controlling for the effects of socially desirable responding, is there still a significant relationship between optimism and life satisfaction scores?

What you need: three continuous variables (e.g. optimism, life satisfaction, socially desirable responding)

Non-parametric alternative: none

Multiple regression

Example of research question: How much of the variance in life satisfaction scores can be explained by the following set of variables: self-esteem, optimism and perceived control? Which of these variables is a better predictor of life satisfaction?

What you need:
- one continuous dependent variable (e.g. life satisfaction); and
- two or more continuous independent variables (e.g. self-esteem, optimism, perceived control).

Diagram:

Self-esteem
Optimism ──────────→ Life satisfaction
Perceived control

Non-parametric alternative: none

Exploring differences between groups

Independent-samples t-test

Example of research question: Are males more optimistic than females?

What you need:
- one categorical independent variable with only *two* groups (e.g. sex: males/females); and
- one continuous dependent variable (e.g. optimism score).

Subjects can belong to only *one group*.

Diagram:

	Males	**Females**
Mean optimism score		

Non-parametric alternative: Mann-Whitney Test

Paired-samples t-test (repeated measures)

Example of research question: Does ten weeks of meditation training result in a decrease in participants' level of anxiety? Is there a change in anxiety levels from Time 1 (pre-intervention) to Time 2 (post-intevention)?

What do you need:
- one categorical independent variable (e.g. Time 1/ Time 2); and
- one continuous dependent variable (e.g. anxiety score).

Same subjects tested on *two* separate occasions: Time 1 (before intervention) and Time 2 (after intervention).

Diagram:

	Time 1	**Time 2**
Mean anxiety score		

Non-parametric alternative: Wilcoxon Signed-Rank Test

One-way between-groups analysis of variance

Example of research question: Is there a difference in optimism scores for people who are under 30, between 31–49 and 50 years and over?

What you need:
- one categorical independent variable with two or more groups (e.g. age: under 30/31–49/50+); and
- one continuous dependent variable (e.g. optimism score).

Diagram:

	Age		
	Under 30	**31–49**	**50 years and over**
Mean optimism score			

Non-parametric alternative: Kruskal-Wallis Test

Two-way between-groups analysis of variance

Example of research question: What is the effect of age on optimism scores for males and females?

What do you need:
- two categorical independent variables (e.g. sex: males/females; age group: under 30/31–49/50+); and
- one continuous dependent variable (e.g. optimism score).

Diagram:

		Age		
		Under 30	**31–49**	**50 years and over**
Mean optimism score	Males			
	Females			

Non-parametric alternative: none

Note: Analysis of variance can also be extended to include three or more independent variables (usually referred to as Factorial Analysis of Variance).

Mixed between-within analysis of variance

Example of research question: Which intervention (maths skills/confidence building) is more effective in reducing participants' fear of statistics, measured across three periods (pre-intervention, post-intervention, three-month follow-up)?

What you need:
- one between-groups independent variable (e.g. type of intervention);
- one within-groups independent variable (e.g. time 1, time 2, time 3); and
- one continuous dependent variable (e.g. scores on Fear of Stats test).

Diagram:

		Time		
		Time 1	**Time 2**	**Time 3**
Mean score on Fear of Statistics test	Maths skills intervention			
	Confidence building intervention			

Non-parametric alternative: none

Multivariate analysis of variance

Example of research question: Are males better adjusted than females in terms of their general physical and psychological health (in terms of anxiety and depression levels and perceived stress)?

What you need: • one categorical independent variable (e.g. sex: males/females); and
• two or more continuous dependent variables (e.g. anxiety, depression, perceived stress).

Diagram:

	Males	**Females**
Anxiety		
Depression		
Perceived stress		

Non-parametric alternative: none

Note: Multivariate analysis of variance can be used with one-way (one independent variable), two-way (two independent variables) and higher-order factorial designs. Covariates can also be included.

Analysis of covariance

Example of research question: Is there a significant difference in the Fear of Statistics test scores for participants in the maths skills group and the confidence building group, while controlling for their pre-test scores on this test?

What do you need: • one categorical independent variable (e.g. type of intervention);
• one continuous dependent variable (e.g. fear of statistics scores at Time 2); and
• one or more continuous covariates (e.g. fear of statistics scores at Time 1).

Non-parametric alternative: none

Note: Analysis of covariance can be used with one-way (one independent variable), two-way (two independent variables) and higher-order factorial designs, and with multivariate designs (two or more dependent variables).

References

The statistical techniques discussed in this chapter are only a small sample of all the different approaches that you can take to data analysis. It is important that you are aware of the existence, and potential uses, of a wide variety of techniques in order to choose the most suitable one for your situation. Some useful readings are suggested below.

If you would like a simple discussion of the major approaches, see:

Cone, J., & Foster, S. (1993). *Dissertations and theses from start to finish*. Washington, DC: American Psychological Association. Chapter 10.

For a coverage of the basic techniques (t-test, analysis of variance, correlation), go back to your basic statistics texts. For example:

Cooper, D. R., & Schindler, P. S. (2003). *Business research methods* (8th edn). Boston: McGraw Hill.
Gravetter, F. J., & Wallnau, L. B. (2000). *Statistics for the behavioral sciences* (5th edn). Belmont, CA: Wadsworth.
Peat, J. (2001). *Health science research: A handbook of quantitative methods*. Sydney: Allen & Unwin.
Runyon, R. P., Coleman, K. A., & Pittenger, D. J. (2000). *Fundamentals of behavioral statistics* (9th edn). Boston: McGraw Hill.

If you are working or studying in the health or medical area a great book, which is also fun to read (trust me!), is:

Norman, G. R., & Streiner, D. L. (2000). *Biostatistics: The bare essentials* (2nd edn). Hamilton: B.C. Decker

If you would like more detailed information, particularly on multivariate statistics, see:

Hair, J. F., Anderson, R. E., Tatham, R. L., & Black, W. C. (1998). *Multivariate data analysis* (5th edn). Upper Saddle River, NJ: Prentice Hall.
Tabachnick, B., & Fidell, L. (2001). *Using multivariate statistics* (4th edn). New York: HarperCollins.

Summary table of the characteristics of the main statistical techniques

Purpose	Example of question	Parametric statistic	Non-parametric alternative	Independent variable	Dependent variable	Essential features
Exploring relationships	What is the relationship between gender and dropout rates from therapy?	None	Chi-square Chapter 22	one categorical variable *Sex: M/F*	one categorical variable *Dropout/complete therapy: Yes/No*	The number of cases in each category is considered, not scores
	Is there a relationship between age and optimism scores?	Pearson product-moment correlation coefficient (r) Chapter 11	Spearman's Rank Order Correlation (rho) Chapter 22	two continuous variables *Age, Optimism scores*		One sample with scores on two different measures, or same measure at Time 1 and Time 2
	After controlling for the effects of socially desirable responding bias, is there still a relationship between optimism and life satisfaction?	Partial correlation Chapter 12	None	two continuous variables and one continuous variable you wish to control for *Optimism, life satisfaction, scores on a social desirability scale*		One sample with scores on two different measures, or same measure at Time 1 and Time 2
	How much of the variance in life satisfaction scores can be explained by self-esteem, perceived control and optimism? Which of these variables is the best predictor?	Multiple regression Chapter 13	None	set of two or more continuous independent variables *Self-esteem, perceived control, optimism*	one continuous dependent variable *Life satisfaction*	One sample with scores on all measures
	What is the underlying structure of the items that make up the Positive and Negative Affect Scale—how many factors are involved?	Factor analysis Chapter 15	None	set of related continuous variables *Items of the Positive and Negative Affect Scale*		One sample, multiple measures
Comparing groups	Are males more likely to dropout of therapy than females?	None	Chi-square Chapter 22	one categorical independent variable *Sex*	one categorical dependent variable *Dropout/complete therapy*	You are interested in the *number of people* in each category, not scores on a scale

Purpose	Example of question	Parametric statistic	Non-parametric alternative	Independent variable	Dependent variable	Essential features
Comparing groups (cont.)	Is there a change in participants' anxiety scores from Time 1 to Time 2?	Paired samples t-test Chapter 16	Wilcoxon Signed-Rank test Chapter 22	one categorical independent variable (two levels) Time 1/Time 2	one continuous dependent variable Anxiety scores	Same people on two different occasions
	Is there a difference in optimism scores for people who are under 35yrs, 36–49yrs and 50+ yrs?	One-way between groups ANOVA Chapter 17	Kruskal-Wallis Chapter 22	one categorical independent variable (three or more levels) Age group	one continuous dependent variable Optimism scores	Three or more groups: different people in each group
	Is there a change in participants' anxiety scores from Time 1, Time 2 and Time 3?	One-way repeated measures ANOVA Chapter 17	Friedman Test Chapter 22	one categorical independent variable (three or more levels) Time 1/Time 2/Time 3	one continuous dependent variable Anxiety scores	Three or more groups: same people on two different occasions
	Is there a difference in the optimism scores for males and females, who are under 35yrs, 36–49yrs and 50+ yrs?	Two-way between groups ANOVA Chapter 18	None	two categorical independent variables (two or more levels) Age group, Sex	one continuous dependent variable Optimism scores	Two or more groups for each independent variable: different people in each group
	Which intervention (maths skills/confidence building) is more effective in reducing participants' fear of statistics, measured across three time periods?	Mixed between-within ANOVA Chapter 19	None	one between-groups independent variable, (two or more levels) one within-groups independent variable (two or more levels) Type of intervention, Time	one continuous dependent variable Fear of Statistics test scores	Two or more groups with different people in each group, each measured on two or more occasions
	Is there a difference between males and females, across three different age groups, in terms of their scores on a variety of adjustment measures (anxiety, depression, and perceived stress)?	Multivariate ANOVA (MANOVA) Chapter 20	None	one or more categorical independent variables (two or more levels) Age group, Sex	two or more related continuous dependent variables Anxiety, depression and perceived stress scores	
	Is there a significant difference in the Fear of Stats test scores for participants in the maths skills group and the confidence building group, while controlling for their scores on this test at Time 1?	Analysis of covariance (ANCOVA) Chapter 21	None	one or more categorical independent variables (two or more levels) one continuous covariate variable Type of intervention, Fear of Stats test scores at Time 1	one continuous dependent variable Fear of Stats test scores at Time 2	

Part Four

Statistical techniques to explore relationships among variables

In the chapters included in this section we will be looking at some of the techniques available in SPSS for exploring relationships among variables. In this section our focus is on detecting and describing relationships among variables. All of the techniques covered here are based on correlation. Correlational techniques are often used by researchers engaged in non-experimental research designs. Unlike experimental designs, variables are not deliberately manipulated or controlled—variables are described as they exist naturally. These techniques can be used to:

- explore the association between pairs of variables (correlation);
- predict scores on one variable from scores on another variable (bivariate regression);
- predict scores on a dependent variable from scores of a number of independent variables (multiple regression); and
- identify the structure underlying a group of related variables (factor analysis).

This family of techniques is used to test models and theories, predict outcomes and assess reliability and validity of scales.

Techniques covered in Part Four

There is a range of techniques available in SPSS to explore relationships. These vary according to the type of research question that needs to be addressed and the types of data available. In this book, however, only the most commonly used techniques are covered.

Correlation (Chapter 11) is used when you wish to describe the strength and direction of the relationship between two variables (usually continuous). It can also be used when one of the variables is dichotomous—that is, it has only two values (e.g. sex: males/females). The statistic obtained is Pearson's product-moment correlation (r). The statistical significance of r is also provided.

Partial correlation (Chapter 12) is used when you wish to explore the relationship between two variables while statistically controlling for a third variable. This is useful when you suspect that the relationship between your two variables of interest may be influenced, or confounded, by the impact of a third variable. Partial correlation statistically removes the influence of the third variable, giving a cleaner picture of the actual relationship between your two variables.

Multiple regression (Chapter 13) allows prediction of a single dependent continuous variable from a group of independent variables. It can be used to test the predictive power of a set of variables and to assess the relative contribution of each individual variable.

Logistic regression (Chapter 14) is used instead of multiple regression when your dependent variable is categorical. It can be used to test the predictive power of a set of variables and to assess the relative contribution of each individual variable.

Factor analysis (Chapter 15) is used when you have a large number of related variables (e.g. the items that make up a scale), and you wish to explore the underlying structure of this set of variables. It is useful in reducing a large number of related variables to a smaller, more manageable, number of dimensions or components.

In the remainder of this introduction to Part Four, I will review some of the basic principles of correlation that are common to all the techniques covered in Part Four. This material should be reviewed before you attempt to use any of the procedures covered in Chapters 11, 12, 13, 14 and 15.

Revision of the basics

Correlation coefficients (e.g. Pearson product-moment correlation) provide a numerical summary of the direction and the strength of the linear relationship between two variables. Pearson correlation coefficients (r) can range from –1 to +1. The sign out the front indicates whether there is a positive correlation (as one variable increases, so too does the other) or a negative correlation (as one variable increases, the other decreases). The size of the absolute value (ignoring the sign) provides information on the strength of the relationship. A perfect correlation of 1 or –1 indicates that the value of one variable can be determined exactly by knowing the value on the other variable. On the other hand, a correlation of 0 indicates no relationship between the two variables. Knowing the value of one of the variables provides no assistance in predicting the value of the second variable.

The relationship between variables can be inspected visually by generating a scatterplot. This is a plot of each pair of scores obtained from the subjects in the sample. Scores on the first variable are plotted along the X (horizontal) axis and the corresponding scores on the second variable plotted on the Y (vertical) axis. An inspection of the scatterplot provides information on both the direction of the relationship (positive or negative) and the strength of the relationship (this is demonstrated in more detail in Chapter 11). A scatterplot of a perfect correlation (r=1 or –1) would show a straight line. A scatterplot when r=0, however, would show a circle of points, with no pattern evident.

Factors to consider when interpreting a correlation coefficient

There are a number of things you need to be careful of when interpreting the results of a correlation analysis, or other techniques based on correlation. Some of the key issues are outlined below, but I would suggest you go back to your statistics books and review this material (see, for example, Gravetter & Wallnau, 2000, pp. 536–540).

Non-linear relationship

The correlation coefficient (e.g. Pearson r) provides an indication of the linear (straight-line) relationship between variables. In situations where the two variables are related in non-linear fashion (e.g. curvilinear), Pearson r will seriously underestimate the strength of the relationship. Always check the scatterplot, particularly if you obtain low values of r.

Outliers

Outliers (values that are substantially lower or higher than the other values in the data set) can have a dramatic effect on the correlation coefficient, particularly in small samples. In some circumstances outliers can make the r value much higher than it should be, and in other circumstances they can result in an underestimate of the true relationship. A scatterplot can be used to check for outliers—just look for values that are sitting out on their own. These could be due to a data entry error (typing 11, instead of 1), a careless answer from a respondent, or it could be a true value from a rather strange individual! If you find an outlier you should check for errors and correct if appropriate. You may also need to consider removing or recoding the offending value, to reduce the effect it is having on the r value (see Chapter 6 for a discussion on outliers).

Restricted range of scores

You should always be careful interpreting correlation coefficients when they come from only a small subsection of the possible range of scores (e.g. using university students to study IQ). Correlation coefficients from studies using a restricted range of cases are often different from studies where the full range of possible scores are sampled. In order to provide an accurate and reliable indicator of the strength of the relationship between two variables there should be as wide a range of scores on each of the two variables as possible. If you are involved in studying extreme groups (e.g. clients with high levels of anxiety) you should not try to generalise any correlation beyond the range of the data used in the sample.

Correlation versus causality

Correlation provides an indication that there is a relationship between two variables; it does not, however, indicate that one variable *causes* the other. The correlation between two variables (A and B) could be due to the fact that A causes B, that B causes A, or (just to complicate matters) that an additional variable (C) causes both A and B. The possibility of a third variable that influences both of your observed variables should always be considered.

To illustrate this point there is the famous story of the strong correlation that one researcher found between ice-cream consumption and the number of homicides reported in New York City. Does eating ice-cream cause people to become violent? No. Both variables (ice-cream consumption and crime rate) were influenced by the weather. During the very hot spells, both the ice-cream consumption and the crime rate increased. Despite the positive correlation obtained, this did not prove that eating ice-cream causes homicidal behaviour. Just as well—the ice-cream manufacturers would very quickly be out of business!

The warning here is clear—watch out for the possible influence of a third, confounding variable when designing your own study. If you suspect the possibility of other variables that might influence your result, see if you can measure these at the same time. By using partial correlation (described in Chapter 12), you can statistically control for these additional variables, and therefore gain a clearer, and less contaminated, indication of the relationship between your two variables of interest.

Statistical versus practical significance

Don't get too excited if your correlation coefficients are 'significant'. With large samples, even quite small correlation coefficients can reach statistical significance. Although statistically significant, the practical significance of a correlation of .2 is very limited. You should focus on the actual size of Pearson's r and the amount of shared variance between the two variables. To interpret the strength of your correlation coefficient you should also take into account other research that has been conducted in your particular topic area. If other researchers in your area have been able to predict only 9 per cent of the variance (a correlation of .3) in a particular outcome (e.g. anxiety), then your study that explains 25 per cent would be impressive in comparison. In other topic areas, 25 per cent of the variance explained may seem small and irrelevant.

Assumptions

There are a number of assumptions common to all the techniques covered in Part Four. These are discussed below. You will need to refer back to these assumptions when performing any of the analyses covered in Chapters 11, 12, 13, 14 and 15.

Level of measurement

The scale of measurement for the variables should be interval or ratio (continuous). The exception to this is if you have one dichotomous independent variable (with only two values: e.g. sex) and one continuous dependent variable. You should, however, have roughly the same number of people or cases in each category of the dichotomous variable.

Related pairs

Each subject must provide a score on both variable X and variable Y (related pairs). Both pieces of information must be from the same subject.

Independence of observations

The observations that make up your data must be independent of one another. That is, each observation or measurement must not be influenced by any other observation or measurement. Violation of this assumption, according to Stevens (1996, p. 238), is very serious. There are a number of research situations that may violate this assumption of independence. Examples of some such studies are described below (these are drawn from Stevens, 1996, p. 239; and Gravetter & Wallnau, 2000, p. 262):

- Studying the performance of students working in pairs or small groups. The behaviour of each member of the group influences all other group members, thereby violating the assumption of independence.
- Studying the TV watching habits and preferences of children drawn from the same family. The behaviour of one child in the family (e.g. watching Program A) is likely to affect all children in that family, therefore the observations are not independent.
- Studying teaching methods within a classroom and examining the impact on students' behaviour and performance. In this situation all students could be influenced by the presence of a small number of trouble-makers, therefore individual behavioural or performance measurements are not independent.

Any situation where the observations or measurements are collected in a group setting, or subjects are involved in some form of interaction with one another, should be considered suspect. In designing your study you should try to ensure that all observations are independent. If you suspect some violation of this assumption, Stevens (1996, p. 241) recommends that you set a more stringent alpha value (e.g. p<.01).

Normality

Scores on each variable should be normally distributed. This can be checked by inspecting the histograms of scores on each variable (see Chapter 6 for instructions).

Linearity

The relationship between the two variables should be linear. This means that when you look at a scatterplot of scores you should see a straight line (roughly), not a curve.

Homoscedasticity

The variability in scores for variable X should be similar at all values of variable Y. Check the scatterplot (see Chapter 6 for instructions). It should show a fairly even cigar shape along its length.

Other issues

Missing data

When you are doing research, particularly with human beings, it is very rare that you will obtain complete data from every case. It is thus important that you inspect your data file for missing data. Run **Descriptives** and find out what percentage of values is missing for each of your variables. If you find a variable with a lot of unexpected missing data you need to ask yourself why. You should also consider whether your missing values are happening randomly, or whether there is some systematic pattern (e.g. lots of women failing to answer the question about their age). SPSS has a **Missing Value Analysis** procedure that may help find patterns in your missing values (see the bottom option under the **Analyze** menu).

You also need to consider how you will deal with missing values when you come to do your statistical analyses. The **Options** button in many of the SPSS statistical procedures offers you choices for how you want SPSS to deal with missing data. It is important that you choose carefully, as it can have dramatic effects on your results. This is particularly important if you are including a list of variables, and repeating the same analysis for all variables (e.g. correlations among a group of variables, t-tests for a series of dependent variables).

- The *Exclude cases listwise* option will include cases in the analysis only if it has full data on *all of the variables* listed in your variables box for that case. A case will be totally excluded from all the analyses if it is missing even one piece of information. This can severely, and unnecessarily, limit your sample size.
- The *Exclude cases pairwise* option, however, excludes the cases (persons) only if they are missing the data required for the specific analysis. They will still be included in any of the analyses for which they have the necessary information.
- The *Replace with mean* option, which is available in some SPSS statistical procedures (e.g. multiple regression), calculates the mean value for the variable

and gives every missing case this value. This option should NEVER be used, as it can severely distort the results of your analysis, particularly if you have a lot of missing values.

Always press the **Options** button for any statistical procedure you conduct and check which of these options is ticked (the default option varies across procedures). I would strongly recommend that you use pairwise exclusion of missing data, unless you have a pressing reason to do otherwise. The only situation where you might need to use listwise exclusion is when you want to refer only to a subset of cases that provided a full set of results.

References

Cohen, J. W. (1988). *Statistical power analysis for the behavioral sciences* (2nd edn). Hillsdale, NJ: Lawrence Erlbaum Associates.

Gravetter, F. J., & Wallnau, L. B. (2000). *Statistics for the behavioral sciences* (5th edn). Belmont, CA: Wadsworth.

Stevens, J. (1996). *Applied multivariate statistics for the social sciences* (3rd edn). Mahway, NJ: Lawrence Erlbaum.

To revise correlation I suggest you go back to your basic statistics books (see list in Recommended references at the end of this book).

For more advanced material, I have listed some titles below (be warned: some of these are pretty heavy-duty!).

Aiken, L. S., & West, S. G. (1991). *Multiple regression: Testing and interpreting interactions*. Newbury Park, CA: Sage.

Berry, W. D. (1993). *Understanding regression assumptions*. Newbury Park, CA: Sage.

Cohen, J., & Cohen, P. (1983). *Applied multiple regression/correlation analysis for the behavioral sciences* (2nd edn). New York: Erlbaum.

Fox, J. (1991). *Regression diagnostics*. Newbury Park, CA: Sage.

Keppel, G., & Zedeck, S. (1989). *Data analysis for research designs: Analysis of variance and multiple regression/correlation approaches*. New York: Freeman.

Stevens, J. (1996). *Applied multivariate statistics for the social sciences* (3rd edn). Mahway, NJ: Lawrence Erlbaum.

Tabachnick, B. G., & Fidell, L. S. (2001). *Using multivariate statistics* (4th edn). New York: HarperCollins.

11 Correlation

Correlation analysis is used to describe the strength and direction of the linear relationship between two variables. There are a number of different statistics available from SPSS, depending on the level of measurement. In this chapter the procedure for obtaining and interpreting a Pearson product-moment correlation coefficient is presented. Pearson product-moment coefficient is designed for interval level (continuous) variables. It can also be used if you have one continuous variable (e.g. scores on a measure of self-esteem) and one dichotomous variable (e.g. sex: M/F). Spearman rank order correlation (designed for use with ordinal level or ranked data) is covered in the chapter on non-parametric techniques (see Chapter 22).

SPSS will calculate two types of correlation for you. First, it will give you a simple bivariate correlation (which just means between two variables), also known as zero-order correlation. SPSS will also allow you to explore the relationship between two variables, while controlling for another variable. This is known as partial correlation. In this chapter the procedure to obtain a bivariate Pearson product-moment correlation coefficient is presented. In Chapter 12 partial correlation is covered.

Pearson correlation coefficients (r) can take on only values from –1 to +1. The sign out the front indicates whether there is a positive correlation (as one variable increases, so too does the other) or a negative correlation (as one variable increases, the other decreases). The size of the absolute value (ignoring the sign) provides an indication of the strength of the relationship. A perfect correlation of 1 or –1 indicates that the value of one variable can be determined exactly by knowing the value on the other variable. A scatterplot of this relationship would show a straight line. On the other hand, a correlation of 0 indicates no relationship between the two variables. Knowing the value on one of the variables provides no assistance in predicting the value on the second variable. A scatterplot would show a circle of points, with no pattern evident.

There are a number of issues associated with the use of correlation. These include the effect of non-linear relationships, outliers, restriction of range, correlation versus causality and statistical versus practical significance. These topics are discussed in the introduction to Part Four of this book. I would strongly recommend that you read through that material before proceeding with the remainder of this chapter.

Details of example

To demonstrate the use of correlation I will explore the interrelationships among some of the variables included in the survey.sav data file provided on the website accompanying this book (see p. xi). The survey was designed to explore the factors that affect respondents' psychological adjustment and wellbeing (see the Appendix for a full description of the study). In this example I am interested in assessing the correlation between respondents' feelings of control and their level of perceived stress. Details of the two variables I will be using are provided in the following table.

File name	Variable name	Variable label	Coding instructions
survey.sav	tpcoiss	Total Perceived Control of Internal States scale	This scale measures the degree to which people feel they have control over their internal states (emotions, thoughts and physical reactions). The total score is calculated by adding all 18 items of the PCOISS scale (pc1 to pc18). Total scores can range from 18 (low levels of control) to 90 (high levels of control).
	tpstress	Total Perceived Stress scale	This scale measures respondents' levels of perceived stress. The total score is calculated by adding all 10 items of the Perceived Stress scale (pss1 to pss10). Total scores can range from 10 (low levels of stress) to 50 (high levels of stress).

If you wish to follow along with this example you should start SPSS and open the survey.sav file. This file can be opened only in SPSS.

Summary for correlation

Example of research question:	Is there a relationship between the amount of control people have over their internal states and their levels of perceived stress? Do people with high levels of perceived control experience lower levels of perceived stress?
What you need:	Two variables: both continuous, or one continuous and the other dichotomous (two values).
What it does:	Correlation describes the relationship between two continuous variables, in terms of both the strength of the relationship and the direction.
Assumptions:	For details of the assumptions for correlation, see the introduction to Part Four.
Non-parametric alternative:	Spearman's Rank Order Correlation (rho); see Chapter 22.

Preliminary analyses for correlation

Before performing a correlation analysis it is a good idea to generate a scatterplot first. This enables you to check for violation of the assumptions of linearity and homoscedasticity (see introduction to Part Four). Inspection of the scatterplots also gives you a better idea of the nature of the relationship between your variables.

Procedure for generating a scatterplot

1. From the menu at the top of the screen click on: **Graphs**, then click on **Scatter**.

2. Click on **Simple**. Click on the **Define** button.

3. Click on the first variable and move it into the **Y-axis** box (this will run vertically). By convention, the dependent variable is usually placed along the Y-axis (e.g. total perceived stress).

4. Click on the second variable and move to the **X-axis** box (this will run across the page). This is usually the independent variable (e.g. total PCOISS).

5. If you would like to add a title, click on **Titles**. Type in a title. Click on **Continue** and then **OK**.

The output generated from this procedure is shown below.

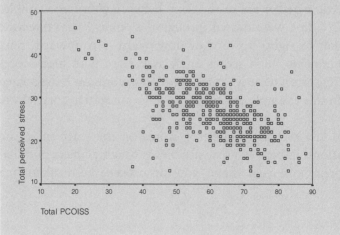

Interpretation of output from scatterplot

The scatterplot can be used to check a number of aspects of the distribution of these two variables.

Step 1: Checking for outliers

Check your scatterplot for outliers—that is, data points that are out on their own, either very high or very low, or away from the main cluster of points.

Extreme outliers are worth checking: were the data entered correctly, could these values be errors? Outliers can seriously influence some analyses, so this is worth investigating. Some statistical texts recommend removing extreme outliers from the data set. Others suggest recoding them down to a value that is not so extreme (see Chapter 6).

Step 2: Inspecting the distribution of data points

The distribution of data points can tell you a number of things:

- Are the data points spread all over the place? This suggests a very low correlation.
- Are all the points neatly arranged in a narrow cigar shape? This suggests quite a strong correlation.
- Could you draw a straight line through the main cluster of points, or would a curved line better represent the points? If a curved line is evident (suggesting a curvilinear relationship), then Pearson correlation should not be used. Pearson correlation assumes a linear relationship.
- What is the shape of the cluster? Is it even from one end to the other? Or does it start off narrow and then get fatter? If this is the case, your data may be violating the assumption of homoscedasticity.

Step 3: Determining the direction of the relationship between the variables

The scatterplot can tell you whether the relationship between your two variables is positive or negative. If a line were drawn through the points, what direction would it point—from left to right; upward or downward? An upward trend indicates a positive relationship, high scores on X associated with high scores on Y. A downward line suggests a negative correlation, low scores on X associated with high scores on Y. In this example we appear to have a negative correlation, of moderate strength.

Once you have explored the distribution of scores on the scatterplot and established that the relationship between the variables is roughly linear, and that the scores are evenly spread in a cigar shape, you can proceed with calculating Pearson's correlation.

Procedure for calculating Pearson product-moment correlation

1. From the menu at the top of the screen click on: **Analyze**, then click on **Correlate**, then on **Bivariate**.

2. Select your two variables and move them into the box marked **Variables** (e.g. total perceived stress, total PCOISS). You can list a whole range of variables here, not just two. In the resulting matrix, the correlation between all possible pairs of variables will be listed. This can be quite large if you list more than just a few variables.

3. Check that the **Pearson** box and the **2 tail** box have a cross in them. The two-tail test of significance means that you are not making any specific prediction concerning the direction of the relationship between the variables (positive/negative). You can choose a **one-tail** test of significance if you have reasons to support a specific direction.

4. Click on the **Options** button.
   For **Missing Values**, click on the **Exclude cases pairwise** box.
   Under **Options** you can also obtain means, standard deviations if you wish. Click on **Continue**.

5. Click **OK**.

The output generated from this procedure is shown below.

Correlations

		Total PCOISS	Total perceived stress
Total PCOISS	Pearson Correlation	1.000	-.581**
	Sig. (2-tailed)	.	.000
	N	431	426
Total perceived stress	Pearson Correlation	-.581**	1.000
	Sig. (2-tailed)	.000	.
	N	426	433

** Correlation is significant at the 0.01 level (2-tailed).

Interpretation of output from correlation

Correlation provides you with a table giving Pearson r correlation coefficients between each pair of variables listed. This can be quite large if you select more than a few variables in your list. For each pair of variables the r value, the significance level and the number of cases is given. There are a number of different aspects of the output that you should consider. We'll look at those steps now.

Step 1: Checking the information about the sample

The first thing to look at in the table labelled **Correlations** is the N (number of cases). Is this correct? If there are a lot of missing data you need to find out why. Did you forget to tick the **Exclude cases pairwise** in the missing data option? Using listwise deletion (the other option), any case with missing data on any of the variables will be removed from the analysis. This can sometimes severely restrict your N. In the above example we have 426 cases that had scores on both of the scales used in this analysis. If a case was missing information on either of these variables it would have been excluded from the analysis.

Step 2: Determining the direction of the relationship

The second thing to consider is the direction of the relationship between the variables. Is there a negative sign in front of the r value? If there is, this means there is a negative correlation between the two variables (i.e. high scores on one are associated with low scores on the other). The interpretation of this depends on the way the variables are scored. Always check with your questionnaire, and remember to take into account that for many scales some items are negatively worded and therefore are reversed before scoring. What do high values really mean? This is one of the major areas of confusion for students, so make sure you get this clear in your mind before you interpret the correlation output. In the example given above the correlation coefficient is negative (–.581), indicating a negative correlation between perceived control and stress. The *more* control people feel they have, the *less* stress they experience.

Step 3: Determining the strength of the relationship

The third thing to consider in the output is the size of the value of Pearson correlation (r). This can range from –1.00 to 1.00. This value will indicate the strength of the relationship between your two variables. A correlation of 0 indicates no relationship at all, a correlation of 1.0 indicates a perfect positive correlation, and a value of –1.0 indicates a perfect negative correlation.

How do you interpret values between 0 and 1? Different authors suggest different interpretations; however, Cohen (1988) suggests the following guidelines:

r=.10 to .29 or r=–.10 to –.29	small
r=.30 to .49 or r=–.30 to –.4.9	medium
r=.50 to 1.0 or r=–.50 to –1.0	large

These guidelines apply whether or not there is a negative sign out the front of your r value. Remember, the negative sign refers only to the direction of the relationship, not the strength. The *strength* of correlation of r=.5 and r=–.5 is the same. It is only in a different *direction*.

In the example presented above there is a large correlation between the two variables (r=–.58), suggesting quite a strong relationship between perceived control and stress.

Step 4: Calculating the coefficient of determination

To get an idea of how much variance your two variables share you can also calculate what is referred to as the coefficient of determination. Sounds impressive, but all you need to do is square your r value (multiply it by itself). To convert this to 'percentage of variance' just multiply by 100 (shift the decimal place two columns to the right). For example, two variables that correlate r=.2 share only .2 × .2 = .04 = 4 per cent of their variance. There is not much overlap between the two variables. A correlation of r=.5, however, means 25 per cent shared variance (.5 × .5 = .25).

In our example the correlation is r=−.581, which when squared indicates 33.76 per cent shared variance. Perceived control helps to explain nearly 34 per cent of the variance in respondents' scores on the perceived stress scale. This is quite a respectable amount of variance explained when compared with a lot of the research conducted in the social sciences.

Step 5: Assessing the significance level

The next thing to consider is the significance level (listed as **Sig. 2 tailed**). This is a very 'messy' area, and should be treated cautiously. The significance of r is strongly influenced by the size of the sample. In a small sample (e.g. N=30), you may have moderate correlations that do not reach statistical significance at the traditional p<.05 level. In large samples (N=100+), however, very small correlations may be statistically significant. Many authors in this area suggest that statistical significance should be reported but ignored, and the focus should be directed at the amount of shared variance (see Step 4).

Presenting the results from correlation

The results of the above example could be presented in a research report as follows:

The relationship between perceived control of internal states (as measured by the PCOISS) and perceived stress (as measured by the Perceived Stress scale) was investigated using Pearson product-moment correlation coefficient. Preliminary analyses were performed to ensure no violation of the assumptions of normality, linearity and homoscedasticity. There was a strong, negative correlation between the two variables [r=−.58, n=426, p<.0005], with high levels of perceived control associated with lower levels of perceived stress.

Correlation is often used to explore the relationship among a group of variables, rather than just two as described above. In this case it would be awkward to report all the individual correlation coefficients in a paragraph; it would be better to present them in a table. One way this could be done is as follows:

TABLE X

Pearson Product-Moment Correlations Between Measures of Perceived Control and Wellbeing

Measures	1	2	3	4
(1) PCOISS				
(2) MAST	.52 ***			
(3) PA	.46 ***	.43 ***		
(4) NA	−.48 ***	−.46 ***	−.29 ***	
(5) LifeSat	.37 ***	.44 ***	.42 ***	−.32 ***

N=428. PCOISS=Perceived Control of Internal States scale; MAST=Mastery scale; PA=Positive Affect scale; NA=Negative Affect scale; LifeSat=Satisfaction with Life scale.

***$p<.001$

Obtaining correlation coefficients between groups of variables

In the previous procedures section I showed you how to obtain correlation coefficients between two continuous variables. If you have a group of variables and you wish to explore the interrelationships among all of them you can ask SPSS to do this all in one procedure. Just include all the variables in the **Variables** box. This can, however, result in an enormous correlation matrix that can be difficult to read and interpret.

Sometimes you want to look at only a subset of all these possible relationships. For example, you might want to look at the relationship between optimism and a number of different adjustment measures (positive affect, negative affect, life satisfaction). You don't want a full correlation matrix, because this would give you correlation coefficients among all the variables, including between each of the various pairs of adjustment measures. There is a way that you can limit the correlation coefficients that are displayed. This involves using what SPSS calls the Syntax Editor (see Figure 11.1).

Figure 11.1

Example of a SPSS
Syntax Editor window

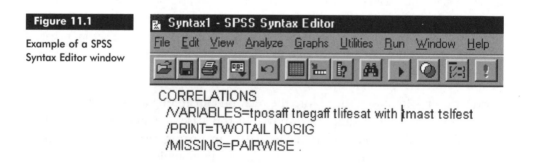

```
CORRELATIONS
 /VARIABLES=tposaff tnegaff tlifesat with tmast tslfest
 /PRINT=TWOTAIL NOSIG
 /MISSING=PAIRWISE .
```

In the good old days, all SPSS commands were given using a special command language or syntax. SPSS still creates these sets of commands to run each of the programs, but all you usually see is the Windows menus, which 'write' the commands for you. Sometimes it is useful to step behind the scenes and write your own commands. This is one such time. The following procedure uses the Syntax Editor to limit the correlation coefficients that are produced by SPSS.

Procedure for obtaining correlation coefficients between one group of variables and another group of variables

1. From the menu at the top of the screen click on: **Analyze**, then click on **Correlation**, then on **Bivariate**.

2. Move the variables of interest into the **Variables box**. List all the first group of variables (e.g. tposaff, tnegaff, tlifesat) followed by the second group (e.g. tpcoiss, tmast).

3. In the output that is generated the first group of variables will appear down the side of the table as rows and the second group will appear across the table as columns.

4. Put your longer list first: this stops your table being too wide to appear on one page.

5. Click on **Paste**. This opens the **Syntax Editor** (see the example of this given in Figure 11.1).

6. Put your cursor between the first group of variables (e.g. tposaff, tnegaff, tlifesat) and the other variables (e.g. tpcoiss and tmast). Type in the word: with. The final syntax should read as follows:

   CORRELATIONS
   /VARIABLES=tposaff tnegaff tlifesat with tpcoiss tmast
   /PRINT=TWOTAIL NOSIG
   /MISSING=PAIRWISE .

This will ask SPSS to calculate correlation coefficients between tmast and tpcoiss and each of the other variables listed.

7. To run this new syntax you need to highlight the text from CORRELATIONS, down to and including the full stop. *It is very important that you include the full stop in the highlighted section.*

8. With this text highlighted, click on the button on the menu bar with a triangle pointing to the right (>) or alternatively click on **Run** from the **Menu**, and then **Selection** from the drop-down menu that appears. This tells SPSS that you wish to run this procedure.

The output generated from this procedure is shown below.

Correlations

		Total PCOISS	Total Mastery
Total Positive Affect	Pearson Correlation	.456**	.432**
	Sig. (2-tailed)	.000	.000
	N	429	436
Total Negative Affect	Pearson Correlation	-.484**	.464**
	Sig. (2-tailed)	.000	.000
	N	428	435
Total Life Satisfaction	Pearson Correlation	.373**	.444**
	Sig. (2-tailed)	.000	.000
	N	429	436

** Correlation is significant at the 0.01 level (2-tailed).

Presented in this manner it is easy to compare the relative strength of the correlations for my two control scales (Total PCOISS, Total mastery) with each of the adjustment measures.

Comparing the correlation coefficients for two groups

Sometimes when doing correlational research you may want to compare the strength of the correlation coefficients for two separate groups. For example, you may want to look at the relationship between optimism and negative affect for males and females separately. One way that you can do this is described now.

Procedure for comparing correlation coefficients for two groups of subjects
A: Split the sample

1. Make sure you have the **Data Editor** window open on the screen in front of you. (If you currently have the Viewer Window open, click on **Window** and choose the **SPSS Data Editor**.)

2. From the menu at the top of the screen click on: **Data**, then click on **Split File**.

3. Click on **Compare Groups**.

4. Move the grouping variable (e.g. sex) into the box labelled **Groups based on**. Click on **OK**.

5. This will split the sample by sex and repeat any analyses that follow for these two groups separately.

B: Correlation

1. Follow the steps in the earlier section of this chapter to request the correlation between your two variables of interest (e.g. Total optimism, Total negative affect). These will be reported separately for the two groups.

Important: Remember, when you have finished looking at males and females separately you will need to turn the **Split File** option off. It stays in place until you specifically turn it off.

To do this, make sure the you have the **Data Editor** window open on the screen in front of you. Click on **Data**, **Split File** and click on the first button: **Analyze all cases, do not create groups**. Click on **OK**.

The output generated from this procedure is shown below.

Correlations

SEX			Total Optimism	Total Negative Affect
MALES	Total Optimism	Pearson Correlation	1.000	-.220**
		Sig. (2-tailed)	.	.003
		N	184	184
	Total Negative Affect	Pearson Correlation	-.220**	1.000
		Sig. (2-tailed)	.003	.
		N	184	185
FEMALES	Total Optimism	Pearson Correlation	1.000	-.394**
		Sig. (2-tailed)	.	.000
		N	251	250
	Total Negative Affect	Pearson Correlation	-.394**	1.000
		Sig. (2-tailed)	.000	.
		N	250	250

** Correlation is significant at the 0.01 level (2-tailed).

Interpretation of output from correlation for two groups

From the output given above, the correlation between Total optimism and Total negative affect for males was r=-.22, while for females it was slightly higher, r=−.39. Although these two values seem different, is this difference big enough to be considered significant? Detailed in the next section is one way that you can test the statistical significance of the difference between these two correlation coefficients. It is important to note that this process is different from testing the statistical significance of the correlation coefficients reported in the output table above. The significance levels reported above (for males: Sig.=.003; for females: Sig.=.000) provide a test of the null hypothesis that the correlation coefficient in the population is 0. The significance test described below, however, assesses the probability that the *difference* in the correlations observed for the two groups (males and females) would occur as a function of a sampling error, when in fact there was no real difference in the strength of the relationship for males and females.

Testing the statistical significance of the difference between correlation coefficients

In this section I will describe the procedure that can be used to find out whether the correlations for the two groups are significantly different. Unfortunately, SPSS will not do this step for you, so it is back to the trusty calculator. This next section might seem rather confusing, but just follow along with the procedure step by step. First we will be converting the r values into z scores, then using an equation to calculate the observed value of z (z obs value). The value obtained will be assessed using some set decision rules to determine the likelihood that the difference in the correlation noted between the two groups could have been due to chance.

Assumptions

As always, there are assumptions to check first. It is assumed that the r values for the two groups were obtained from random samples and that the two groups of cases are independent (not the same subjects tested twice). The distribution of scores for the two groups is assumed to be normal (see histograms for the two groups). It is also necessary to have at least 20 cases in each of the groups.

Step 1: Convert each of the r values into z values

The first step in the comparison process is to convert the two r values that you obtained into a standard score form (referred to as z scores). This is done for a number of mathematical reasons, primarily to ensure that the sampling distributions are approximately normal.

From the SPSS output find the r value and N for Group 1 (males) and Group 2 (females).

Males: $r_1 = -.22$ Females: $r_2 = -.394$
$N_1 = 184$ $N_2 = 250$

Using Table 11.1, find the z value that corresponds with each of the r values.
Males: $z_1 = -.224$ Females: $z_2 = -.418$

r	z_r	r	z_r	r	z_r	r	z_r	r	z_r
.000	.000	.200	.203	.400	.424	.600	.693	.800	1.099
.005	.005	.205	.208	.405	.430	.605	.701	.805	1.113
.010	.010	.210	.213	.410	.436	.610	.709	.810	1.127
.015	.015	.215	.218	.415	.442	.615	.717	.815	1.142
.020	.020	.220	.224	.420	.448	.620	.725	.820	1.157
.025	.025	.225	.229	.425	.454	.625	.733	.825	1.172
.030	.030	.230	.234	.430	.460	.630	.741	.830	1.188
.035	.035	.235	.239	.435	.466	.636	.750	.835	1.204
.040	.040	.240	.245	.440	.472	.640	.758	.840	1.221
.045	.045	.245	.250	.445	.478	.645	.767	.845	1.238
.050	.050	.250	.255	.450	.485	.650	.775	.850	1.256
.055	.055	.255	.261	.455	.491	.655	.784	.855	1.274
.060	.060	.260	.266	.460	.497	.660	.793	.860	1.293
.065	.065	.265	.271	.465	.504	.665	.802	.865	1.313
.070	.070	.270	.277	.470	.510	.670	.811	.870	1.333
.075	.075	.275	.282	.475	.517	.675	.820	.875	1.354
.080	.080	.280	.288	.460	.523	.680	.829	.880	1.376
.085	.085	.285	.293	.485	.530	.685	.838	.885	1.398
.090	.090	.290	.299	.490	.536	.690	.848	.890	1.422
.095	.095	.295	.304	.495	.543	.695	.858	.895	1.447
.100	.100	.300	.310	.500	.549	.700	.867	.900	1.472
.105	.105	.305	.315	.505	.556	.705	.877	.905	1.499
.110	.110	.310	.321	.510	.563	.710	.887	.910	1.528
.115	.116	.315	.326	.515	.570	.715	.897	.915	1.557
.120	.121	.320	.332	.520	.576	.720	.908	.920	1.589
.125	.126	.325	.337	.525	.583	.725	.918	.925	1.623
.130	.131	.330	.343	.530	.590	.730	.929	.930	1.658
.135	.136	.335	.348	.535	.597	.735	.940	.935	1.697
.140	.141	.340	.354	.540	.604	.740	.950	.940	1.738
.145	.146	.345	.360	.545	.611	.745	.962	.945	1.783
.150	.151	.350	.365	.550	.618	.750	.973	.950	1.832
.155	.156	.355	.371	.555	.626	.755	.984	.955	1.886
.160	.161	.360	.377	.560	.633	.760	.996	.960	1.946
.165	.167	.365	.383	.565	.640	.765	1.008	.965	2.014
.170	.172	.370	.388	.570	.648	.770	1.020	.970	2.092
.175	.177	.375	.394	.575	.655	.775	1.033	.975	2.185
.180	.182	.380	.400	.580	.662	.780	1.045	.980	2.298
.185	.187	.385	.406	.585	.670	.785	1.058	.985	2.443
.190	.192	.390	.412	.590	.678	.790	1.071	.990	2.647
.195	.198	.395	.418	.595	.685	.795	1.085	.995	2.994

Table 11.1

Tranformation of r to z

Source: McCall (1990); originally from Edwards, A. L. (1967). *Statistical methods* (2nd edition). Holt, Rinehart and Winston.

Step 2: Put these values into the equation to calculate z_{obs}

The equation that you need to use is given below. Just slot your values into the required spots and then crunch away on your calculator (it helps if you have set out your information clearly and neatly).

$$z_{obs} = \frac{z_1 - z_2}{\sqrt{\dfrac{1}{N_1 - 3} + \dfrac{1}{N_2 - 3}}}$$

$$z_{obs} = \frac{.224 - .418}{\sqrt{\dfrac{1}{184 - 3} + \dfrac{1}{250 - 3}}}$$

$$z_{obs} = \frac{-.194}{\sqrt{\dfrac{1}{181} + \dfrac{1}{247}}}$$

$$z_{obs} = \frac{-.194}{\sqrt{.0055 + .004}}$$

$$z_{obs} = \frac{-.194}{\sqrt{.0095}}$$

$$z_{obs} = \frac{-.194}{.0975} = -1.99$$

Step 3: Determine if the z_{obs} value is statistically significant

If the z_{obs} value that you obtained is between –1.96 and +1.96 then you *cannot* say that there is a statistically significant difference between the two correlation coefficients.

In statistical language, you are able to reject the null hypothesis (no difference between the two groups) *only* if your z value is outside of these two boundaries.

The decision rule therefore is:

If $-1.96 < z_{obs} < 1.96$: correlation coefficients are *not* statistically significantly different.

If $z_{obs} \leq -1.96$ or $z_{obs} \geq 1.96$: coefficients are statistically significantly different.

In the example calculated above for males and females we obtained a z_{obs} value of –1.99. This is outside the specified bounds, so we can conclude that there is a statistically significant difference in the strength of the correlation between optimism and negative affect for males and females. Optimism explains significantly more of the variance in negative affect for females than for males.

Additional exercises

Health

Data file: *sleep.sav*. See Appendix for details of the data file.

1. Check the strength of the correlation between scores on the Sleepiness and Associated Sensations Scale (*totSAS*) and the Epworth Sleepiness Scale (*ess*).

2. Use Syntax to assess the correlations between the Epworth Sleepiness Scale (*ess*) and each of the individual items that make up the Sleepiness and Associated Sensations Scale (*fatigue, lethargy, tired, sleepy, energy*).

Reference

Cohen, J. W. (1988). *Statistical power analysis for the behavioral sciences* (2nd edn). Hillsdale, NJ: Lawrence Erlbaum Associates.

12 Partial correlation

Partial correlation is similar to Pearson product-moment correlation (described in Chapter 11), except that it allows you to control for an additional variable. This is usually a variable that you suspect might be influencing your two variables of interest. By statistically removing the influence of this confounding variable, you can get a clearer and more accurate indication of the relationship between your two variables.

In the introduction to Part Four the influence of contaminating or confounding variables was discussed (see section on correlation versus causality). This occurs when the relationship between two variables (A and B) is influenced, at least to some extent, by a third variable (C). This can serve to artificially inflate the size of the correlation coefficient obtained. This relationship can be represented graphically as:

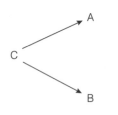

In this case, A and B may look as if they are related, but in fact their apparent relationship is due, to a large extent, to the influence of C. If you were to statistically control for the variable C, then the correlation between A and B is likely to be reduced, resulting in a smaller correlation coefficient.

Details of example

To illustrate the use of partial correlation I will use the same statistical example as described in Chapter 11, but extend the analysis further to control for an additional variable. This time I am interested in exploring the relationship between scores on the Perceived Control of Internal States scale (PCOISS) and scores on the Perceived Stress scale, while controlling for what is known as socially desirable responding bias. This variable simply refers to people's tendency to present themselves in a positive or 'socially desirable' way (also known as 'faking good') when completing questionnaires. This tendency is measured by the Marlowe–Crowne Social

Desirability scale (Crowne & Marlowe, 1960). A short version of this scale (Strahan & Gerbasi, 1972) was included in the questionnaire used to measure the other two variables. The abbreviation used as a variable name in the data file was tmarlow. If you would like to follow along with the example presented below you should start SPSS and open the file labelled survey.sav which is included on the website accompanying this book (see p. xi). This data file can only be opened in SPSS.

Summary for partial correlation

Example of research question: After controlling for subjects' tendency to present themselves in a positive light on self-report scales, is there still a significant relationship between perceived control of internal states (PCOISS) and levels of perceived stress?

What you need: • three variables: all continuous;
• two variables that you wish to explore the relationship between (e.g. total PCOISS, total perceived stress); and
• one variable that you wish to control for (e.g. total social desirability: tmarlow).

What it does: Very simply, partial correlation allows you to explore the relationship between two variables, while statistically controlling for (getting rid of) the effect of another variable that you think might be contaminating or influencing the relationship.

Assumptions: For full details of the assumptions for correlation, see the introduction to Part Four.

Procedure for partial correlation

1. From the menu at the top of the screen click on: **Analyze**, then click on **Correlate**, then on **Partial**.

2. Click on the two continuous variables that you want to correlate (e.g. total PCOISS, total perceived stress). Click on the arrow to move these into the **Variables** box.

3. Click on the variable that you wish to control for (e.g. tmarlow, which measures social desirability). Move into the **Controlling for** box.

4. Choose whether you want one-tail or two-tail significance (two-tail was used in this example).

5. Click on **Options**.
   • In the **Missing Values** section click on **Exclude cases pairwise**.
   • In the **Statistics** section click on **Zero Order correlations**.

6. Click on **Continue** and then **OK**.

The output generated from this procedure is shown below.

Correlations

Control Variables			total PCOISS	total perceived stress	total social desirability
-none-[a]	total PCOISS	Correlation	1.000	-.581	.295
		Significance (2-tailed)	.	.000	.000
		df	0	424	425
	total perceived stress	Correlation	-.581	1.000	-.228
		Significance (2-tailed)	.000	.	.000
		df	424	0	426
	total social desirability	Correlation	.295	-.228	1.000
		Significance (2-tailed)	.000	.000	.
		df	425	426	0
total social desirability	total PCOISS	Correlation	1.000	-.552	
		Significance (2-tailed)	.	.000	
		df	0	423	
	total perceived stress	Correlation	-.552	1.000	
		Significance (2-tailed)	.000	.	
		df	423	0	

a. Cells contain zero-order (Pearson) correlations.

Interpretation of output from partial correlation

The output provides you with a table made up of two sections:

1. In the top half of the table is the normal Pearson product-moment correlation matrix between your two variables of interest (e.g. perceived control and perceived stress), *not* controlling for your other variable. In this case the correlation is –.58. The word 'none' in the left-hand column indicates that no control variable is in operation.
2. The bottom half of the table repeats the same set of correlation analyses, but this time controlling for (taking out) the effects of your control variable (e.g. social desirability). In this case the new partial correlation is –.55.

You should compare these two sets of correlation coefficients to see whether controlling for the additional variable had any impact on the relationship between your two variables. In this example there was only a small decrease in the strength of the correlation (from –.58 to –.55). This suggests that the observed relationship between perceived control and perceived stress is not due merely to the influence of socially desirable responding.

Presenting the results from partial correlation

An example of how to report the above results is:

Partial correlation was used to explore the relationship between perceived control of internal states (as measured by the PCOISS) and perceived stress (measured by the Perceived Stress scale), while controlling for scores on the Marlowe–Crowne Social Desirability scale. Preliminary analyses were performed to ensure no violation of the assumptions of normality, linearity and homoscedasticity. There was a strong, positive, partial correlation between perceived control of internal states and perceived stress [$r=-.55$, $n=423$, $p<.0005$], with high levels of perceived control being associated with lower levels of perceived stress. An inspection of the zero order correlation ($r=-.58$) suggested that controlling for socially desirable responding had very little effect on the strength of the relationship between these two variables.

Additional exercises

Health

Data file: *sleep.sav*. See Appendix for details of the data file.

1. Check the strength of the correlation between scores on the Sleepiness and Associated Sensations Scale (*totSAS*) and the impact of sleep problems on overall wellbeing (*impact6*) while controlling for *age*. Compare the zero order correlation (Pearson Correlation) and the partial correlation coefficient. Does controlling for *age* make a difference?

References

Crowne, D. P., & Marlowe, D. (1960). A new scale of social desirability independent of psychopathology. *Journal of Consulting Psychology, 24*, 349–354.

Strahan, R., & Gerbasi, K. (1972). Short, homogeneous version of the Marlowe–Crowne Social Desirability Scale. *Journal of Clinical Psychology, 28*, 191–193.

13 Multiple regression

In this chapter I will briefly outline how to use SPSS for Windows to run multiple regression analyses. This is a *very* simplified outline. It is important that you do more reading on multiple regression before using it in your own research. A good reference is Chapter 5 in Tabachanick and Fiddell (2001), which covers the underlying theory, the different types of multiple regression analyses and the assumptions that you need to check.

Multiple regression is not just one technique but a family of techniques that can be used to explore the relationship between one continuous dependent variable and a number of independent variables or predictors (usually continuous). Multiple regression is based on correlation (covered in Chapter 11), but allows a more sophisticated exploration of the interrelationship among a set of variables. This makes it ideal for the investigation of more complex real-life, rather than laboratory-based, research questions. However, you cannot just throw variables into a multiple regression and hope that, magically, answers will appear. You should have a sound theoretical or conceptual reason for the analysis and, in particular, the order of variables entering the equation. Don't use multiple regression as a fishing expedition.

Multiple regression can be used to address a variety of research questions. It can tell you how well a set of variables is able to predict a particular outcome. For example, you may be interested in exploring how well a set of subscales on an intelligence test is able to predict performance on a specific task. Multiple regression will provide you with information about the model as a whole (all subscales), and the relative contribution of each of the variables that make up the model (individual subscales). As an extension of this, multiple regression will allow you to test whether adding a variable (e.g. motivation) contributes to the predictive ability of the model, over and above those variables already included in the model. Multiple regression can also be used to statistically control for an additional variable (or variables) when exploring the predictive ability of the model. Some of the main types of research questions that multiple regression can be used to address are:

- how well a *set of variables* is able to predict a particular outcome;
- *which variable* in a set of variables is the best predictor of an outcome; and
- whether a particular predictor variable is still able to predict an outcome when the effects of another variable are *controlled for* (e.g. socially desirable responding).

Major types of multiple regression

There are a number of different types of multiple regression analyses that you can use, depending on the nature of the question you wish to address. The three main types of multiple regression analyses are:

- standard or simultaneous;
- hierarchical or sequential; and
- stepwise.

Typical of the statistical literature, you will find different authors using different terms when describing these three main types of multiple regression—very confusing for an experienced researcher, let alone a beginner to the area!

Standard multiple regression

In standard multiple regression all the independent (or predictor) variables are entered into the equation simultaneously. Each independent variable is evaluated in terms of its predictive power, over and above that offered by all the other independent variables. This is the most commonly used multiple regression analysis. You would use this approach if you had a set of variables (e.g. various personality scales) and wanted to know how much variance in a dependent variable (e.g. anxiety) they were able to explain as a group or block. This approach would also tell you how much unique variance in the dependent variable each of the independent variables explained.

Hierarchical multiple regression

In hierarchical regression (also called sequential) the independent variables are entered into the equation in the order specified by the researcher based on theoretical grounds. Variables or sets of variables are entered in steps (or blocks), with each independent variable being assessed in terms of what it adds to the prediction of the dependent variable, after the previous variables have been controlled for. For example, if you wanted to know how well optimism predicts life satisfaction, after the effect of age is controlled for, you would enter age in Block 1 and then Optimism in Block 2. Once all sets of variables are entered, the overall model is assessed in terms of its ability to predict the dependent measure. The relative contribution of each block of variables is also assessed.

Stepwise multiple regression

In stepwise regression the researcher provides SPSS with a list of independent variables and then allows the program to select which variables it will enter, and in which order they go into the equation, based on a set of statistical criteria. There are three different versions of this approach: forward selection, backward deletion and stepwise regression. There are a number of problems with these approaches, and some controversy in the literature concerning their use (and abuse). Before using these approaches I would strongly recommend that you read up on the issues involved (see p. 138 in Tabachnick & Fidell, 2001). It is important that you understand what is involved, how to choose the appropriate variables and how to interpret the output that you receive.

Assumptions of multiple regression

Multiple regression is one of the fussier of the statistical techniques. It makes a number of assumptions about the data, and it is not all that forgiving if they are violated. It is not the technique to use on small samples, where the distribution of scores is very skewed! The following summary of the major assumptions is taken from Chapter 5, Tabachnick and Fidell (2001). It would be a good idea to read this chapter before proceeding with your analysis. You should also review the material covered in the introduction to Part Four of this book, which covers the basics of correlation, and see the list of Recommended references at the back of the book. The SPSS procedures for testing these assumptions are discussed in more detail in the examples provided later in this chapter.

Sample size

The issue at stake here is generalisability. That is, with small samples you may obtain a result that does not generalise (cannot be repeated) with other samples. If your results do not generalise to other samples, then they are of little scientific value. So how many cases or subjects do you need? Different authors tend to give different guidelines concerning the number of cases required for multiple regression. Stevens (1996, p. 72) recommends that 'for social science research, about 15 subjects per predictor are needed for a reliable equation'. Tabachnick and Fidell (2001, p. 117) give a formula for calculating sample size requirements, taking into account the number of independent variables that you wish to use: $N > 50 + 8m$ (where m = number of independent variables). If you have five independent variables you will need 90 cases. More cases are needed if the dependent variable is skewed. For stepwise regression there should be a ratio of 40 cases for every independent variable.

Multicollinearity and singularity

This refers to the relationship among the independent variables. Multicollinearity exists when the independent variables are highly correlated (r=.9 and above).

Singularity occurs when one independent variable is actually a combination of other independent variables (e.g. when both subscale scores and the total score of a scale are included). Multiple regression doesn't like multicollinearity or singularity, and these certainly don't contribute to a good regression model, so always check for these problems before you start.

Outliers

Multiple regression is very sensitive to outliers (very high or very low scores). Checking for extreme scores should be part of the initial data screening process (see Chapter 6). You should do this for all the variables, both dependent and independent, that you will be using in your regression analysis. Outliers can either be deleted from the data set or, alternatively, given a score for that variable that is high, but not too different from the remaining cluster of scores.

Additional procedures for detecting outliers are also included in the multiple regression program. Outliers on your dependent variable can be identified from the standardised residual plot that can be requested (described in the example presented later in this chapter). Tabachnick and Fidell (2001, p. 122) define outliers as those with standardised residual values above about 3.3 (or less than –3.3).

Normality, linearity, homoscedasticity, independence of residuals

These all refer to various aspects of the distribution of scores and the nature of the underlying relationship between the variables. These assumptions can be checked from the *residuals* scatterplots which are generated as part of the multiple regression procedure. Residuals are the differences between the obtained and the predicted dependent variable (DV) scores. The residuals scatterplots allow you to check:

- *normality:* the residuals should be normally distributed about the predicted DV scores;
- *linearity:* the residuals should have a straight-line relationship with predicted DV scores; and
- *homoscedasticity:* the variance of the residuals about predicted DV scores should be the same for all predicted scores.

The interpretation of the residuals scatterplots generated by SPSS is discussed later in this chapter; however, for a more detailed discussion of this rather complex topic, see Chapter 5 in Tabachnick and Fidell (2001).

Further reading on multiple regression can be found in the Recommended references at the end of this book.

Details of example

To illustrate the use of multiple regression I will be using a series of examples taken from the survey.sav data file that is included on the website with this book (see p. xi). The survey was designed to explore the factors that affect respondents' psychological adjustment and wellbeing (see the Appendix for full details of the study). For the multiple regression example detailed below I will be exploring the impact of respondents' perceptions of control on their levels of perceived stress. The literature in this area suggests that if people feel that they are in control of their lives, they are less likely to experience 'stress'. In the questionnaire there were two different measures of control (see the Appendix for the references for these scales). These include the Mastery scale, which measures the degree to which people feel they have control over the events in their lives; and the Perceived Control of Internal States Scale (PCOISS), which measures the degree to which people feel they have control over their internal states (their emotions, thoughts and physical reactions).

In this example I am interested in exploring how well the Mastery scale and the PCOISS are able to predict scores on a measure of perceived stress. The variables used in the examples covered in this chapter are presented in the following table.

File name	Variable name	Variable label	Coding instructions
survey.sav	tpstress	Total Perceived Stress	Total score on the Perceived Stress scale. High scores indicate high levels of stress.
	tpcoiss	Total Perceived Control of Internal States	Total score on the Perceived Control of Internal States scale. High scores indicate greater control over internal states.
	tmast	Total Mastery	Total score on the Mastery scale. High scores indicate higher levels of perceived control over events and circumstances.
	tmarlow	Total Social Desirability	Total scores on the Marlowe–Crowne Social Desirability scale which measures the degree to which people try to present themselves in a positive light.
	age	Age	Age in years.

It is a good idea to work through these examples on the computer using this data file. Hands-on practice is always better than just reading about it in a book. Feel free to 'play' with the data file—substitute other variables for the ones that were used in the example. See what results you get, and try to interpret them.

Doing is the best way of learning—particularly with these more complex techniques. Practice will help build your confidence for when you come to analyse your own data.

The examples included below cover only the use of Standard Multiple Regression and Hierarchical Regression. Because of the criticism that has been levelled at the use of Stepwise Multiple Regression techniques, these approaches will not be illustrated here. If you are desperate to use these techniques (despite the warnings) I suggest you read Tabachnick and Fidell (2001) or one of the other texts listed in the References at the end of the chapter.

Summary for multiple regression

Example of research questions:	1. How well do the two measures of control (mastery, PCOISS) predict perceived stress? How much variance in perceived stress scores can be explained by scores on these two scales? 2. Which is the best predictor of perceived stress: control of external events (Mastery scale), or control of internal states (PCOISS)? 3. If we control for the possible effect of age and socially desirable responding, is this set of variables still able to predict a significant amount of the variance in perceived stress?
What you need:	• One continuous dependent variable (total perceived stress); and • Two or more continuous independent variables (mastery, PCOISS). (You can also use dichotomous independent variables, e.g. males=1, females=2.)
What it does:	Multiple regression tells you how much of the variance in your dependent variable can be explained by your independent variables. It also gives you an indication of the relative contribution of each independent variable. Tests allow you to determine the statistical significance of the results, both in terms of the model itself, and the individual independent variables.
Assumptions:	The major assumptions for multiple regression are described in an earlier section of this chapter. Some of these assumptions can be checked as part of the multiple regression analysis (these are illustrated in the example that follows).

Standard multiple regression

In this example two questions will be addressed:

Question 1: How well do the two measures of control (mastery, PCOISS) predict perceived stress? How much variance in perceived stress scores can be explained by scores on these two scales?

Question 2: Which is the best predictor of perceived stress: control of external events (Mastery scale) or control of internal states (PCOISS)?

To explore these questions I will be using standard multiple regression. This involves all of the independent variables being entered into the equation at once. The results will indicate how well this set of variables is able to predict stress levels; and it will also tell us how much *unique* variance each of the independent variables (mastery, PCOISS) explains in the dependent variable, *over and above* the other independent variables included in the set.

Procedure for standard multiple regression

1. From the menu at the top of the screen click on: **Analyze**, then click on **Regression**, then on **Linear**.

2. Click on your continuous dependent variable (e.g. total perceived stress: tpstress) and move it into the **Dependent** box.

3. Click on your independent variables (total mastery: tmast; total PCOISS: tpcoiss) and move them into the **Independent** box.

4. For **Method**, make sure **Enter** is selected (this will give you standard multiple regression).

5. Click on the **Statistics** button.
    *   Tick the box marked **Estimates, Confidence Intervals, Model fit, Descriptives, Part and partial correlations** and **Collinearity diagnostics**.
    *   In the **Residuals** section tick the **Casewise diagnostics** and **Outliers outside 3 standard deviations**.
    *   Click on **Continue**.

6. Click on the **Options** button. In the **Missing Values** section click on **Exclude cases pairwise**.

7. Click on the **Plots** button.
   • Click on *****ZRESID** and the arrow button to move this into the **Y** box.
   • Click on *****ZPRED** and the arrow button to move this into the **X** box.
   • In the section headed **Standardized Residual Plots**, tick the **Normal probability plot** option.
   • Click on **Continue**.

8. Click on the **Save** button.
   • In the section labelled **Distances** tick the **Mahalanobis** box (this will identify multivariate outliers for you) and **Cook's**.
   • Click on **Continue**.

9. Click on **OK**.

The output generated from this procedure is shown below.

Correlations

		Total perceived stress	Total Mastery	Total PCOISS
Pearson Correlation	Total perceived stress	1.000	-.612	-.581
	Total Mastery	-.612	1.000	.521
	Total PCOISS	-.581	.521	1.000
Sig. (1-tailed)	Total perceived stress	.	.000	.000
	Total Mastery	.000	.	.000
	Total PCOISS	.000	.000	.
N	Total perceived stress	433	433	426
	Total Mastery	433	436	429
	Total PCOISS	426	429	431

Model Summary [b]

Model	R	R Square	Adjusted R Square	Std. Error of the Estimate
1	.684[a]	.468	.466	4.27

a. Predictors: (Constant), Total PCOISS, Total Mastery

b. Dependent Variable: Total perceived stress

ANOVA[b]

Model		Sum of Squares	df	Mean Square	F	Sig.
1	Regression	6806.728	2	3403.364	186.341	.000[a]
	Residual	7725.756	423	18.264		
	Total	14532.484	425			

a. Predictors: (Constant), Total PCOISS, Total Mastery

b. Dependent Variable: Total perceived stress

Coefficients[a]

Model		Unstandardized Coefficients		Standardized Coefficients	t	Sig.	95% Confidence Interval for B		Correlations			Collinearity Statistics	
		B	Std. Error	Beta			Lower Bound	Upper Bound	Zero order	Partial	Part	Tolerance	VIF
1	(Constant)	50.971	1.273		40.035	.000	48.469	53.474					
	tmast total mastery	-.625	.061	-.424	-10.222	.000	-.745	-.505	-.612	-.445	-.362	.729	1.372
	tpcoiss total PCOISS	-.175	.020	-.360	-8.660	.000	-.215	-.136	-.581	-.388	-.307	.729	1.372

a. Dependent Variable: tpstress total perceived stress

Casewise Diagnostics [a]

Case Number	Std. Residual	Total perceived stress	Predicted Value	Residual
152	-3.473	14	28.84	-14.84

a. Dependent Variable: Total perceived stress

Residuals Statistics [a]

	Minimum	Maximum	Mean	Std. Deviation	N
Predicted Value	18.02	41.30	26.73	4.00	429
Std. Predicted Value	-2.176	3.642	.001	1.000	429
Standard Error of Predicted Value	.21	.80	.34	.11	429
Adjusted Predicted Value	18.04	41.38	26.75	4.01	426
Residual	-14.84	12.62	3.35E-03	4.27	426
Std. Residual	-3.473	2.952	.001	.999	426
Stud. Residual	-3.513	2.970	.001	1.003	426
Deleted Residual	-15.18	12.77	4.18E-03	4.31	426
Stud. Deleted Residual	-3.561	2.998	.001	1.006	426
Mahal. Distance	.004	13.905	1.993	2.233	429
Cook's Distance	.000	.094	.003	.008	426
Centered Leverage Value	.000	.033	.005	.005	429

a. Dependent Variable: Total perceived stress

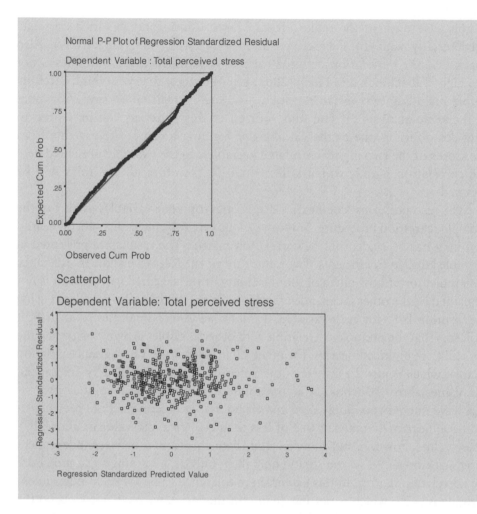

Interpretation of output from standard multiple regression

As with the output from most of the SPSS procedures, there are lots of rather confusing numbers generated as output from regression. To help you make sense of this, I will take you on a guided tour of the output that was obtained to answer Question 1.

Step 1: Checking the assumptions

Multicollinearity. The correlations between the variables in your model are provided in the table labelled **Correlations**. Check that your independent variables show at least some relationship with your dependent variable (above .3 preferably).

In this case both of the scales (Total Mastery and Total PCOISS) correlate substantially with Total Perceived Stress (–.61 and –.58 respectively). Also check that the correlation between each of your independent variables is not too high. Tabachnick and Fidell (2001, p. 84) suggest that you 'think carefully before including two variables with a bivariate correlation of, say, .7 or more in the same analysis'. If you find yourself in this situation you may need to consider omitting one of the variables or forming a composite variable from the scores of the two highly correlated variables. In the example presented here the correlation is .52, which is less than .7, therefore all variables will be retained.

SPSS also performs 'collinearity diagnostics' on your variables as part of the multiple regression procedure. This can pick up on problems with multicollinearity that may not be evident in the correlation matrix. The results are presented in the table labelled Coefficients. Two values are given: Tolerance and VIF. Tolerance is an indicator of how much of the variability of the specified independent is not explained by the other independent variables in the model and is calculated using the formula $1-R^2$ for each variable. If this value is very small (less than .10), it indicates that the multiple correlation with other variables is high, suggesting the possibility of multicollinearity. The other value given is the VIF (Variance inflation factor), which is just the inverse of the Tolerance value (1 divided by Tolerance). VIF values above 10 would be a concern here, indicating multicollinearity.

I have quoted commonly used cut-off points for determining the presence of multicollinearity (tolerance value of less than .10, or a VIF value of above 10). These values, however, still allow for quite high correlations between independent variables (above .9), so you should take them only as a warning sign, and check the correlation matrix. In this example the tolerance value for each independent variable is .729, which is not less than .10; therefore, we have not violated the multicollinearity assumption. This is also supported by the VIF value, which is 1.372, which is well below the cut-off of 10. These results are not surprising, given that the Pearson's correlation coefficient between these two independent variables was only .52 (see Correlations table).

If you exceed these values in your own results, you should seriously consider removing one of the highly intercorrelated independent variables from the model.

Outliers, Normality, Linearity, Homoscedasticity, Independence of Residuals. One of the ways that these assumptions can be checked is by inspecting the residuals scatterplot and the Normal Probability Plot of the regression standardised residuals that were requested as part of the analysis. These are presented at the end of the output. In the Normal Probability Plot you are hoping that your points will lie in a reasonably straight diagonal line from bottom left to top right. This would suggest no major deviations from normality. In the Scatterplot of the standardised residuals (the second plot displayed) you are hoping that the residuals will be roughly rectangularly distributed, with most of the scores concentrated in the centre (along the 0 point). What you don't want to see is a clear or systematic pattern to your residuals (e.g. curvilinear, or higher on one side than the other). Deviations from a centralised rectangle suggest some violation of the assumptions. If you find this to be the case with your data, I suggest you see Tabachnick and Fidell (2001, pp. 154–157) for a full description of how to interpret a residuals plot and how to assess the impact that violations may have on your analysis.

The presence of outliers can also be detected from the **Scatterplot**. Tabachnick and Fidell (2001) define outliers as cases that have a standardised residual (as displayed in the scatterplot) of more than 3.3 or less than –3.3. With large samples, it is not uncommon to find a number of outlying residuals. If you find only a few, it may not be necessary to take any action.

Outliers can also be checked by inspecting the Mahalanobis distances that are produced by the multiple regression program. These do not appear in the output, but instead are presented in the data file as an extra variable at the end of your data file (Mah_1). To identify which cases are outliers you will need to determine the critical chi-square value, using the number of independent variables as the degrees of freedom. A full list of these values can be obtained from any statistics text (see Tabachnick & Fidell, 2001, Table C.4). Tabachnick and Fidell suggest using an alpha level of .001. Using Tabachnick and Fidell's (2001) guidelines I have summarised some of the key values for you in Table 13.1.

Number of indep. variables	Critical value	Number of indep. variables	Critical value	Number of indep. variables	Critical value
2	13.82	4	18.47	6	22.46
3	16.27	5	20.52	7	24.32

Table 13.1

Critical values for evaluating Mahalanobis distance values

Source: Extracted and adapted from a table in Tabachnik and Fidell (1996); originally from Pearson, E. S. and Hartley, H. O. (Eds) (1958). *Biometrika tables for statisticians* (vol. 1, 2nd edn). New York: Cambridge University Press.

To use this table you need to:

*   determine how many independent variables will be included in your multiple regression analysis;
*   find this value in one of the shaded columns; and
*   read across to the adjacent column to identify the appropriate critical value.

In this example I have two independent variables; therefore the critical value is 13.82. To find out if any of the cases have a Mahalanobis distance value exceeding this value I need to run an additional analysis. This is a similar procedure to that detailed in Chapter 6 to identify outliers (using **Descriptives, Explore** and requesting **Outliers** from the list of **Statistics**). This time, though, the variable to use is Mah_1 (this is the extra variable created by SPSS). The five highest values will be displayed; check that none of the values exceed the critical value obtained from the table above. If you ask for the program to label the cases by ID then you will be able to identify any case that does exceed this value. Although I have not displayed it here, when I checked the Mahalanobis distances using the above procedure I found one outlying case (ID number 507, with a value of 17.77). Given the size of the data file, it is not unusual for a few outliers to appear, so I will not worry too much about this one case.

The other information in the output concerning unusual cases is in the Table titled **Casewise Diagnostics**. This presents information about cases that have standardised residual values above 3.0 or below −3.0. In a normally distributed sample we would expect only 1 per cent of cases to fall outside this range. In this sample we have found one case (case number 359) with a residual value of −3.475. You can see from the **Casewise Diagnostics** table that this person recorded a total perceived stress score of 14, but our model predicted a value of 28.85. Clearly our model did not predict this person's score very well—they are much less stressed than we predicted. To check whether this strange case is having any undue influence on the results for our model as a whole, we can check the value for **Cook's Distance** given towards the bottom of the **Residuals Statistics** table. According to Tabachnick and Fidell (2001, p. 69), cases with values larger than 1 are a potential problem. In our example the maximum value for Cook's Distance is .09, suggesting no major problems. In your own data, if you obtain a maximum value above 1 you will need to go back to your data file, sort cases by the new variable that SPSS created at the end of your file (COO_1), which is the Cook's Distance values for each case. Check each of the cases with values above 1—you may need to consider removing the offending case/cases.

Step 2: Evaluating the model

Look in the **Model Summary** box and check the value given under the heading **R Square**. This tells you how much of the variance in the dependent variable (perceived stress) is explained by the model (which includes the variables of Total Mastery and Total PCOISS). In this case the value is .468. Expressed as a percentage (multiply by 100, by shifting the decimal point two places to the right), this means that our model (which includes Mastery and PCOISS) explains 46.8 per cent of the variance in perceived stress. This is quite a respectable result (particularly when you compare it to some of the results that are reported in the journals!).

You will notice that SPSS also provides an **Adjusted R Square** value in the output. When a small sample is involved, the R square value in the sample tends to be a rather optimistic overestimation of the true value in the population (see Tabachnick & Fidell, 2001, p. 147). The Adjusted R square statistic 'corrects' this value to provide a better estimate of the true population value. If you have a small sample you may wish to consider reporting this value, rather than the normal R Square value.

To assess the statistical significance of the result it is necessary to look in the table labelled ANOVA. This tests the null hypothesis that multiple R in the population equals 0. The model in this example reaches statistical significance (Sig = .000, this really means p<.0005).

Step 3: Evaluating each of the independent variables

The next thing we want to know is which of the variables included in the model contributed to the prediction of the dependent variable. We find this information in the output box labelled **Coefficients**. Look in the column labelled **Beta** under **Standardised Coefficients**. To compare the different variables it is important that you look at the *standardised* coefficients, not the *unstandardised* ones. 'Standardised' means that these values for each of the different variables have been converted to the same scale so that you can compare them. If you were interested in constructing a regression equation, you would use the unstandardised coefficient values listed as B.

In this case we are interested in *comparing* the contribution of each independent variable; therefore we will use the beta values. Look down the Beta column and find which beta value is the largest (ignoring any negative signs out the front). In this case the largest beta coefficient is –.424, which is for Total Mastery. This means that this variable makes the strongest unique contribution to explaining the dependent variable, when the variance explained by all other variables in the model is controlled for. The Beta value for Total PCOISS was slightly lower (–.36), indicating that it made less of a contribution.

For each of these variables, check the value in the column marked Sig. This tells you whether this variable is making a statistically significant *unique* contribution to the equation. This is very dependent on which variables are included in the equation, and how much overlap there is among the independent variables. If the **Sig.** value is less than .05 (.01, .0001, etc.), then the variable is making a significant unique contribution to the prediction of the dependent variable. If greater than .05, then you can conclude that that variable is not making a significant unique contribution to the prediction of your dependent variable. This may be due to overlap with other independent variables in the model. In this case, both Total Mastery and Total PCOISS made a unique, and statistically significant, contribution to the prediction of perceived stress scores.

The other potentially useful piece of information in the coefficients table is the **Part** correlation coefficients. Just to confuse matters, you will also see these coefficients referred to as semipartial correlation coefficients (see Tabachnick and Fidell, 2001, p. 140). If you square this value (whatever it is called) you get an indication of the contribution of that variable to the total R squared. In other words, it tells you how much of the total variance in the dependent variable is uniquely explained by that variable and how much R squared would drop if it wasn't included in your model. In this example the Mastery scale has a part correlation coefficient of –.36. If we square this (multiply it by itself) we get .13, indicating that Mastery uniquely explains 13 per cent of the variance in Total Perceived Stress scores. For the PCOISS the value is –.307, which squared gives us .09, indicating a unique contribution of 9 per cent to the explanation of variance in perceived stress.

Note that the total R squared value for the model (in this case .466, or 46 per cent explained variance) does not equal all the squared part correlation values added up (.13+.09=.22). This is because the part correlation values represent only the unique contribution of each variable, with any overlap or shared variance removed or partialled out. The total R squared value, however, includes the unique variance explained by each variable and also that shared. In this case the two independent variables are reasonably strongly correlated (r=.52 as shown in the Correlations table); therefore there is a lot of shared variance that is statistically removed when they are both included in the model.

The results of the analyses presented above allow us to answer the two questions posed at the beginning of this section. Our model, which includes control of external events (Mastery) and control of internal states (PCOISS), explains 46.8 per cent of the variance in perceived stress (Question 1). Of these two variables, mastery makes the largest unique contribution (beta=–.42), although PCOISS also made a statistically significant contribution (beta=–.36) (Question 2).

The beta values obtained in this analysis can also be used for other more practical purposes than the theoretical model testing shown here. Standardised beta values indicate the number of standard deviations that scores in the dependent variable would change if there was a one standard deviation unit change in the predictor. In the current example, if we could increase Mastery scores by one standard deviation (which is 3.97, from the Descriptive Statistics table), then the perceived stress scores would be likely to drop by .42 standard deviation units. If we multiplied this value by 5.85 (the standard deviation of perceived stress scores), we would get .42 × 5.85 = 2.46. This can be useful information, particularly when used in business settings. For example, you may want to know how much

Strange looking numbers

In your output you might come across strange looking numbers which take the form: 1.24E–02. These numbers have been written in scientific notation. They can easily be converted to a 'normal' number for presentation in your results. The number after the E tells you how many places you need to shift the decimal point. If the number after the E is negative, shift the decimal point to the left; if it is positive, shift it to the right. Therefore, 1.24E–02 becomes 0.124. The number 1.24E02 would be written as 124.

You can prevent these small numbers from being displayed in this strange way. Go to **Edit, Options** and put a tick in the box labelled: **No scientific notation for small numbers in tables**. Rerun the analysis and normal numbers should appear in all your output.

your sales figures are likely to increase if you put more money into advertising. Set up a model with sales figures as the dependent variable and advertising expenditure and other relevant variables as independent variables, and then use the beta values as shown above.

Hierarchical multiple regression

To illustrate the use of hierarchical multiple regression I will address a question that follows on from that discussed in the previous example. This time I will evaluate the ability of the model (which includes Total Mastery and Total PCOISS) to predict perceived stress scores, after controlling for a number of additional variables (age, social desirability). The question is as follows:

Question 3: If we control for the possible effect of age and socially desirable responding, is our set of variables (Mastery, PCOISS) still able to predict a significant amount of the variance in perceived stress?

To address this question we will be using hierarchical multiple regression (also referred to as sequential regression). This means that we will be entering our variables in steps or blocks in a predetermined order (not letting the computer decide, as would be the case for stepwise regression). In the first block we will 'force' age and socially desirable responding into the analysis. This has the effect of statistically controlling for these variables.

In the second step we enter the other independent variables into the 'equation' as a block, just as we did in the previous example. The difference this time is that the possible effect of age and socially desirable responding has been 'removed' and we can then see whether our block of independent variables are still able to explain some of the remaining variance in our dependent variable.

Procedure for hierarchical multiple regression

1. From the menu at the top of the screen click on: **Analyze**, then click on **Regression**, then on **Linear**.

2. Choose your continuous dependent variable (e.g. total perceived stress) and move it into the **Dependent** box.

3. Move the variables you wish to control for into the **Independent** box (e.g. age, total social desirability). This will be the first block of variables to be entered in the analysis (Block 1 of 1).

4. Click on the button marked **Next**. This will give you a second independent variables box to enter your second block of variables into (you should see Block 2 of 2).

5. Choose your next block of independent variables (e.g. total mastery, Total PCOISS).

6. In the **Method** box make sure that this is set to the default (**Enter**). This will give you standard multiple regression for each block of variables entered.

7. Click on the **Statistics** button. Tick the boxes marked **Estimates**, **Model fit**, **R squared change**, **Descriptives**, **Part and partial correlations** and **Collinearity diagnostics**. Click on **Continue**.

8. Click on the **Options** button. In the **Missing Values** section click on **Exclude cases pairwise**.

9. Click on the **Save** button. Click on **Mahalonobis** and **Cook's**. Click on **Continue** and then **OK**.

Some of the output generated from this procedure is shown below.

Model Summary ^c

Model	R	R Square	Adjusted R Square	Std. Error of the Estimate	R Square Change	F Change	df1	df2	Sig. F Change
1	.238ª	.057	.052	5.69	.057	12.711	2	423	.000
2	.688ᵇ	.474	.469	4.26	.417	166.873	2	421	.000

a. Predictors: (Constant), AGE, Total social desirability
b. Predictors: (Constant), AGE, Total social desirability, Total Mastery, Total PCOISS
c. Dependent Variable: Total perceived stress

ANOVA ^c

Model		Sum of Squares	df	Mean Square	F	Sig.
1	Regression	823.865	2	411.932	12.711	.000ª
	Residual	13708.620	423	32.408		
	Total	14532.484	425			
2	Regression	6885.760	4	1721.440	94.776	.000ᵇ
	Residual	7646.724	421	18.163		
	Total	14532.484	425			

a. Predictors: (Constant), AGE, Total social desirability
b. Predictors: (Constant), AGE, Total social desirability, Total Mastery, Total PCOISS
c. Dependent Variable: Total perceived stress

Coefficients^a

Model		Unstandardized Coefficients		Standardized Coefficients	t	Sig.	Correlations			Collinearity Statistics	
		B	Std. Error	Beta			Zero-order	Partial	Part	Tolerance	VIF
1	(Constant)	31.076	.983		31.605	.000					
	total social desirability	-.599	.140	-.209	-4.271	.000	-.228	-.203	-.202	.928	1.08
	age	-.031	.022	-.070	-1.438	.151	-.127	-.070	-.068	.928	1.08
2	(Constant)	51.922	1.366		38.008	.000					
	total social desirability	-.149	.108	-.052	-1.373	.171	-.228	-.067	-.049	.871	1.15
	age	-.021	.017	-.047	-1.239	.216	-.127	-.060	-.044	.860	1.16
	total mastery	-.641	.062	-.435	-10.286	.000	-.612	-.448	-.364	.699	1.43
	total PCOISS	-.160	.022	-.327	-7.373	.000	-.581	-.338	-.261	.635	1.57

a. Dependent Variable: total perceived stress

Interpretation of output from hierarchical multiple regression

The output generated from this analysis is similar to the previous output, but with some extra pieces of information. In the **Model Summary** box there are two models listed. Model 1 refers to the first block of variables that were entered (Total social desirability and age), while Model 2 includes all the variables that were entered in both blocks (Total social desirability, age, Total mastery, Total PCOISS).

Step 1: Evaluating the model

Check the **R Square** values in the first Model summary box. After the variables in Block 1 (social desirability, age) have been entered, the overall model explains 5.7 per cent of the variance ($.057 \times 100$). After Block 2 variables (Total Mastery, Total PCOISS) have also been included, the model *as a whole* explains 47.4 per cent ($.474 \times 100$). It is important to note that this second R square value includes all the variables from both blocks, not just those included in the second step.

To find out how much of this overall variance is explained by our variables of interest (Mastery, POCISS) after the effects of age and socially desirable responding are removed, you need to look in the column labelled **R Square change**. In the output presented above you will see, on the line marked Model 2, that the R square change value is .417. This means that Mastery and PCOISS explain an *additional* 41.7 per cent ($.417 \times 100$) of the variance in perceived stress, even when the effects of age and socially desirable responding are statistically controlled for. This is a statistically significant contribution, as indicated by the Sig. F change value for this line (.000). The ANOVA table indicates that the model as a whole (which includes both blocks of variables) is significant [$F (4, 421)=94.78$, $p<.0005$).

Step 2: Evaluating each of the independent variables

To find out how well each of the variables contributes to the equation we need to look in the **Coefficients** table. Always look in the Model 2 row. This summarises the results, with *all* the variables entered into the equation. Scanning the **Sig.** column, there are only two variables that make a statistically significant contribution (less than .05). In order of importance they are: Mastery (beta=-.44, after rounding up) and Total PCOISS (beta=-.33, after rounding). Neither age nor social desirability made a unique contribution. Remember, these beta values represent the unique contribution of each variable, when the overlapping effects of all other variables are statistically removed. In different equations, with a different set of independent variables, or with a different sample, these values would change.

Presenting the results from multiple regression

There are a number of different ways of presenting the results of multiple regression, depending on the type of analysis conducted and the nature of the research question. I would recommend that you see Tabachnick and Fidell (2001, Chapter 5, pp. 163–164) for a full discussion of the presentation of multiple regression analyses. These authors provide examples of standard and sequential multiple regression results in the level of detail required for a thesis.

The *Publication Manual* of the American Psychological Association also provides guidelines and an example of how to present the results of multiple regression (1994, pp. 130–132). As a minimum it suggests that you should indicate what type of analysis was performed (standard or hierarchical), provide both standardised (beta) and unstandardised (B) coefficients (with their standard errors: SE) and, if you performed a hierarchical multiple regression, that you should also provide the R square change values for each step. Associated probability values should also be provided.

It would be a good idea to look for examples of the presentation of different statistical analysis in the journals relevant to your topic area. Different journals have different requirements and expectations. Given the severe space limitations in journals these days, often only a brief summary of the results is presented and readers are encouraged to contact the author for a copy of the full results.

Additional exercises

Health

Data file: *sleep.sav.* See Appendix for details of the data file.

1. Conduct a standard multiple regression to explore factors that impact on people's level of daytime sleepiness. For your dependent variable use the Sleepiness and Associated Sensations Scale total score (*totSAS*). For independent variables use *sex*, *age*, physical fitness rating (*fitrate*) and scores on the HADS Depression scale (*depress*). Assess how

much of the variance in total sleepiness scores is explained by the set of variables (check your R-squared value). Which of the variables make a unique significant contribution (check your beta values)?

2. Repeat the above analysis, but this time use a hierarchical multiple regression procedure entering sex and age in the first block of variables, and physical fitness and depression scores in the second block. After controlling for the demographic variables of sex and age, do the other two predictor variables make a significant contribution to explaining variance in sleepiness scores? How much additional variance in sleepiness is explained by physical fitness and depression, after controlling for sex and age?

References

Stevens, J. (1996). *Applied multivariate statistics for the social sciences* (3rd edn). Mahway, NJ: Lawrence Erlbaum.

Tabachnick, B. G., & Fidell, L. S. (2001). *Using multivariate statistics* (4th edn). New York: HarperCollins.

Multiple regression

Aiken, L. S., & West, S. G. (1991). *Multiple regression: Testing and interpreting interactions*. Newbury Park, CA: Sage.

Berry, W. D. (1993). *Understanding regression assumptions*. Newbury Park, CA: Sage.

Cohen, J., & Cohen, P. (1983). *Applied multiple regression/correlation analysis for the behavioral sciences* (2nd edn). New York: Erlbaum.

Fox, J. (1991). *Regression Diagnostics*. Newbury Park, CA: Sage.

Keppel, G., & Zedeck, S. (1989). *Data analysis for research designs: Analysis of variance and multiple regression/correlation approaches*. New York: Freeman.

Stevens, J. (1996). *Applied multivariate statistics for the social sciences* (3rd edn). Mahway, NJ: Lawrence Erlbaum.

Tabachnick, B. G., & Fidell, L. S. (2001). *Using multivariate statistics* (4th edn). New York: HarperCollins.

14 Logistic regression

In Chapter 13, on multiple regression, we explored a technique to assess the impact of a set of predictors on a dependent variable (perceived stress). In that case the dependent variable was measured as a continuous variable (with scores ranging from 10 to 50). There are many research situations, however, when the dependent variable of interest is categorical (e.g. win/lose; fail/pass; dead/alive). Unfortunately, multiple regression is not suitable when you have categorical dependent variables. For multiple regression your dependent variable (the thing that you are trying to explain or predict) needs to be a continuous variable, with scores reasonably normally distributed.

Logistic regression allows you to test models to predict categorical outcomes with two or more categories. Your predictor (independent) variables can be either categorical or continuous, or a mix of both in the one model. There is a family of logistic regression techniques available in SPSS that will allow you to explore the predictive ability of sets or blocks of variables, and to specify the entry of variables. The purpose of this example is to demonstrate just the basics of logistic regression. I will therefore be using a Forced Entry Method, which is the default procedure available in SPSS. In this approach all predictor variables are tested in one block to assess their predictive ability, while controlling for the effects of other predictors in the model. Other techniques—for example, the stepwise procedures (e.g. forward and backward)—allow you to specify a large group of potential predictors from which SPSS can pick a subset that provides the best predictive power. These stepwise procedures have been criticised (in both logistic and multiple regression) because they can be heavily influenced by random variation in the data, with variables being included or removed from the model on purely statistical grounds (see discussion in Tabachnick & Fidell, 2001, p. 535).

In this chapter I will demonstrate how to use SPSS to perform logistic regression with a dichotomous dependent variable (i.e. with only two categories or values). Here we will use the SPSS procedure labelled Binary Logistic. If your dependent variable has more than two categories, you will need to use the Multinomial Logistic set of procedures (not covered here, but available in SPSS—see the Help menu). Logistic regression is a complex technique, and I would strongly recommend further reading if you intend to use it (some suggested readings are listed at the end of this chapter).

Assumptions

Sample size

As with most statistical techniques, you need to consider the size and nature of your sample if you intend to use logistic regression. One of the issues concerns the number of cases you have in your sample and the number of predictors (independent variables) you wish to include in your model. If you have a small sample with a large number of predictors you may have problems with the analysis (including the problem of the solution failing to converge). This is particularly a problem when you have categorical predictors with limited cases in each category. Always run Descriptive Statistics on each of your predictors, and consider collapsing or deleting categories if they have limited numbers.

Multicollinearity

As discussed in Chapter 13, you should always check for high inter-correlations among your predictor (independent) variables. Ideally, your predictor variables will be strongly related to your dependent variable but not strongly related to each other. Unfortunately, there is no formal way in the logistic regression procedure of SPSS to test for multicollinearity, but you can choose to use the procedure described in Chapter 13 to request collinearity diagnostics under the Statistics button. Ignore the rest of the output but focus on the Coefficients table, and the columns labelled Collinearity Statistics. Tolerance values that are very low (less than .1) indicate that the variable has high correlations with other variables in the model. You may need to reconsider the set of variables that you wish to include in the model, and remove one of the highly inter-correlating variables.

Outliers

It is important to check for the presence of outliers, or cases that are not well explained by your model. In logistic regression terms, a case may be strongly predicted by your model to be one category but in reality be classified in the other category. These outlying cases can be identified by inspecting the residuals, a particularly important step if you have problems with the goodness of fit of your model.

Details of example

To demonstrate the use of logistic regression I will be using a real data file (sleep.sav) available from the website for this book (see web address on p. xi). These data were obtained from a survey I conducted on a sample of university staff to identify the prevalence of sleep-related problems and their impact (see Appendix X). In the survey, respondents were asked whether they considered that they had a sleep-related problem (yes/no). This variable will be used as the dependent variable in this analysis. The set of predictors (independent variables) includes sex, age, the number of hours of sleep the person gets per weeknight, whether they have trouble falling asleep, and whether they have difficulty staying asleep.

Each of the variables was subjected to recoding of their original scores to ensure their suitability for this analysis. The categorical variables were recoded from their original coding so that 0=no and 1=yes.

File name	Variable name	Variable label	Coding instructions
Sleep.sav	probsleeprec	Prob sleep recode	0=no, 1=yes
	Sex	sex	0=female, 1=male
	age	Age	Age in years
	hourwnit	hrs sleep per week night	hours of sleep per week night
	getsleprec	Prob fall asleep rec	Do you have trouble falling asleep? Score recoded to: 0=no, 1=yes
	stayslprec	Prob stay asleep rec	Do you have trouble staying asleep? Score recoded to: 0=no, 1=yes

Data preparation: coding of responses

In order to make sense of the results of logistic regression it is important that you set up the coding of responses to each of your variables carefully. For the dichotomous dependent variable you should code the responses as 0 and 1 (or recode existing values using the Recode procedure in SPSS—see Chapter 8). The value of 0 should be assigned to whichever response indicates a lack or absence of the characteristic of interest. In this example 0 is used to code the answer No to the question 'Do you have a problem with your sleep?'. The value of 1 is used to indicate a Yes answer. A similar approach is used when coding the independent variables. Here we have coded the answer Yes as 1, for both of the categorical variables relating to difficulty getting to sleep and difficulty staying asleep. For continuous independent variables (number of hours sleep per night), high values should indicate more of the characteristic of interest.

Summary for logistic regression

Example of research question: What factors predict the likelihood that respondents would report that they had a problem with their sleep?

What you need:
- One categorical (dichotomous) dependent variable (problem with sleep: No/Yes, coded 0/1); and
- Two or more continuous or categorical predictor (independent) variables. Code dichotomous variables using 0 and 1 (e.g. sex, trouble getting to sleep, trouble staying asleep). Measure continuous variables so that high values indicate more of the characteristic of interest (e.g. age, hours of sleep per night).

What it does: Logistic regression allows you to assess how well your set of predictor variables predicts or explains your categorical dependent variable. It gives you an indication of the adequacy of your model (set of predictor variables) by assessing 'goodness of fit'. It provides an indication of the relative importance of each predictor variable, or the interaction among your predictor variables. It provides a summary of the accuracy of the classification of cases based on the mode, allowing the calculation of the sensitivity and specificity of the model and the positive and negative predictive values.

Assumptions: Logistic regression does not make assumptions concerning the distribution of scores for the predictor variables; however, it is sensitive to high correlations among the predictor variables. This is referred to as multicollinearity. Outliers can also influence the results of logistic regression.

Procedure for logistic regression

- From the menu at the top of the screen click on: **Analyze**, then click on **Regression** and then **Binary Logistic**.

- Choose your categorical dependent variable (e.g. probsleeprec) and move it into the **Dependent** box.

- Click on your predictor variables (sex, age, getsleprec, stayslprec, hourwnit) and move them into the box labelled **Covariates**.

- For **Method**, make sure that **Enter** is displayed (this is the SPSS default, and forces all variables into the model in one block).

- If you have any categorical predictors you will need to click on the **Categorical** button at the bottom of the dialogue box. Highlight each of the categorical variables (sex, getsleprec, stayslprec) and move them into the **Categorical covariates** box.

- Highlight each of your categorical variables in turn and click on the button labelled **First** in the **Change contrast** section. Click on the **Change** button and you will see the word (first) appear after the variable. This will set the group to be used as the reference as the first group listed. Click on **Continue**.

- Click on the **Options** button. Select **Classification plots**, **Hosmer-Lemeshow goodness of fit**, **Casewise listing of residuals**, **Iteration history** and **CI for Exp(B)**. Click on **Continue** and then **OK**.

Some of the output generated from this procedure is shown below.

Logistic regression

Case Processing Summary

Unweighted Cases[a]		N	Percent
Selected Cases	Included in Analysis	241	88.9
	Missing Cases	30	11.1
	Total	271	100.0
Unselected Cases		0	.0
Total		271	100.0

a. If weight is in effect, see classification table for the total number of cases.

Dependent Variable Encoding

Original Value	Internal Value
no	0
yes	1

Categorical Variables Codings

		Frequency	Paramete (1)
prob stay asleep rec	no	138	.000
	yes	103	1.000
prob fall asleep rec	no	151	.000
	yes	90	1.000
sex	female	140	.000
	male	101	1.000

Block 0: Beginning Block

Iteration History[a,b,c]

Iteration		-2 Log likelihood	Coefficients Constant
Step 0	1	328.996	-.290
	2	328.996	-.293
	3	328.996	-.293

a. Constant is included in the model.

b. Initial -2 Log Likelihood: 328.996

c. Estimation terminated at iteration number 3 because parameter estimates changed by less than .001.

Classification Table[a,b]

			Predicted		
			prob sleep recode 01		Percentage Correct
	Observed		no	yes	
Step 0	prob sleep recode 01	no	138	0	100.0
		yes	103	0	.0
	Overall Percentage				57.3

a. Constant is included in the model.

b. The cut value is .500

Variables in the Equation

		B	S.E.	Wald	df	Sig.	Exp(B)
Step 0	Constant	-.293	.130	5.047	1	.025	.746

Variables not in the Equation

			Score	df	Sig.
Step 0	Variables	sex(1)	1.209	1	.272
		age	.795	1	.373
		getsleprec(1)	19.812	1	.000
		stayslprec(1)	58.183	1	.000
		hourwnit	17.709	1	.000
	Overall Statistics		70.017	5	.000

Block 1: Method = Enter

Omnibus Tests of Model Coefficients

		Chi-square	df	Sig.
Step 1	Step	76.020	5	.000
	Block	76.020	5	.000
	Model	76.020	5	.000

Model Summary

Step	-2 Log likelihood	Cox & Snell R Square	Nagelkerke R Square
1	252.976[a]	.271	.363

a. Estimation terminated at iteration number 5 because parameter estimates changed by less than .001.

Hosmer and Lemeshow Test

Step	Chi-square	df	Sig.
1	10.019	8	.264

Classification Table[a]

			Predicted		
			prob sleep recode 01		Percentage Correct
Observed			no	yes	
Step 1	prob sleep recode 01	no	110	28	79.7
		yes	32	71	68.9
	Overall Percentage				75.1

a. The cut value is .500

Variables in the Equation

		B	S.E.	Wald	df	Sig.	Exp(B)	95.0% C.I.for EXP(B) Lower	Upper
Step 1[a]	sex(1)	-.108	.315	.118	1	.731	.897	.484	1.663
	age	-.006	.014	.193	1	.660	.994	.968	1.021
	getsleprec(1)	.716	.339	4.464	1	.035	2.046	1.053	3.976
	stayslprec(1)	1.984	.325	37.311	1	.000	7.274	3.848	13.748
	hourwnit	-.448	.165	7.366	1	.007	.639	.462	.883
	Constant	1.953	1.451	1.812	1	.178	7.053		

a. Variable(s) entered on step 1: sex, age, getsleprec, stayslprec, hourwnit.

Casewise List[b]

		Observed			Temporary Variable	
Case	Selected Status[a]	prob sleep recode 01	Predicted	Predicted Group	Resid	ZResid
42	S	n**	.870	y	-.870	-2.583
224	S	y**	.126	n	.874	2.633
227	S	y**	.133	n	.867	2.554
235	S	y**	.119	n	.881	2.721
265	S	y**	.121	n	.879	2.697

a. S = Selected, U = Unselected cases, and ** = Misclassified cases.

b. Cases with studentized residuals greater than 2.000 are listed.

Interpretion of output from logistic regression

As with most SPSS output, there is an almost overwhelming amount of information provided from logistic regression. I will highlight only the key aspects.

The first thing to check is the details concerning sample size provided in the **Case Processing Summary** table. Make sure you have the number of cases that you expect.

The next table, **Dependent Variable Encoding**, tells you how SPSS has dealt with the coding of your dependent variable (in this case whether people consider they have a problem with their sleep). SPSS needs the variables to be coded using 0 and 1, but it will do it for you if your own coding does not match this (e.g. if your values are 1 and 2). We wanted to make sure that our 'problem with sleep' variable was coded so that 0=no problem and 1=problem, so we created a new variable (using the Recode procedure—see Chapter 8), recoding the original response of 1=yes, 2=no to the format preferred by SPSS of 1=yes, 0=no. This coding of the existence of the problem being represented by 1 and the lack of problem as 0 makes the interpretation of the output a little easier.

Check the coding of your independent (predictor) variables in the next table, labelled **Categorical Variables Codings**. Also check the number of cases you have in each category in the column headed **Frequency**.

The next section of the output, headed **Block 0**, is the results of the analysis without any of our independent variables used in the model. This will serve as a baseline later for comparing the model with our predictor variables included. In the **Classification** table the overall percentage of correctly classified cases is 57.3 per cent. In this case SPSS classified (guessed) that all cases would not have a problem with their sleep (only because there was a higher percentage of people answering No to the question). We hope that later, when our set of predictor variables is entered, we will be able to improve the accuracy of these predictions.

Skip down to the next section, headed **Block 1**. This is where our model (set of predictor variables) is tested.

The **Omnibus Tests of Model Coefficients** gives us an overall indication of how well the model performs, over and above the results obtained for Block 0, with none of the predictors entered into the model. This is referred to as a 'goodness of fit' test. For this set of results we want a highly significant value (the **Sig.** value should be less than .05). In this case the value is .000 (which really means p<.0005). Therefore, the model (with our set of variables used as predictors) is better than SPSS's original guess shown in Block 0, which assumed that everyone would report no problem with their sleep. The chi-square value, which we will need to report in our results, is 76.02 with 5 degrees of freedom.

The results shown in the table headed **Hosmer and Lemeshow Test** also supports our model as being worthwhile. This test, which SPSS states is the most reliable test of model fit available in SPSS, is interpreted very differently from the omnibus test discussed above. For the Hosmer-Lemeshow Goodness of Fit Test poor fit is indicated by a significance value less than .05, so to support our model we actually want a value greater than .05. In our example the chi-square value for the Hosmer-Lemeshow Test is 10.019 with a significance level of .264. This value is larger than .05, therefore indicating support for the model.

The table headed **Model Summary** gives us another piece of information about the usefulness of the model. The Cox & Snell R Square and the Nagelkerke R Square values provide an indication of the amount of variation in the dependent variable explained by the model (from a minimum value of 0 to a maximum of approximately 1). These are described as pseudo R square statistics, rather than the true R square values that you will see provided in the multiple regression output. In this example the two values are .271 and .363, suggesting that between 27.1 per cent and 36.3 per cent of the variability is explained by this set of variables.

The next table in the output to consider is the **Classification table**. This provides us with an indication of how well the model is able to predict the correct category (sleep problem/no sleep problem) for each case. We can compare this with the

Classification table shown for Block 0, to see how much improvement there is when the predictor variables are included in the model. The model correctly classified 75.1 per cent of cases overall (sometimes referred to as the percentage accuracy in classification: PAC), an improvement over the 57.3 per cent in Block 0.

The results displayed in this table can also be used to calculate the additional statistics that you often see reported in the medical literature. The sensitivity of the model is the percentage of the group that has the characteristic of interest (e.g. sleep problem) that has been accurately identified by the model (the true positives). In this example we were able to correctly classify 68.9 per cent of the people who did have a sleep problem. The specificity of the model is the percentage of the group without the characteristic of interest (no sleep problem) that is correctly identified (true negatives). In this example the specificity is 79.7 per cent (people without a sleep problem correctly predicted not to have a sleep problem by the model). The positive predictive value is the percentage of cases that the model classifies as having the characteristic that is actually observed in this group. To calculate this for the current example you need to divide the number of cases in the predicted=yes, observed=yes cell (71) by the total number in the predicted=yes cells (28+71=99) and multiply by 100 to give a percentage. This gives us 71 divided by 99 × 100 = 71.7 per cent. Therefore the positive predictive value is 71.7 per cent, indicating that of the people predicted to have a sleep problem our model accurately picked 71.7 per cent of them. The negative predictive value is the percentage of cases predicted by the model not to have the characteristic that is actually observed not to have the characteristic. In the current example the necessary values from the classification table are: 110 divided by (110+32) × 100 = 77.5 per cent. For further information on the use of classification tables, see Wright (1995, p. 229), or for a simple worked example, see Peat (2001, p. 237).

The **Variables in the Equation** table gives us information about the contribution or importance of each of our predictor variables. The test that is used here is known as the Wald test, and you will see the value of the statistic for each predictor in the column labelled Wald. Scan down the column labelled Sig. looking for values less than .05. These are the variables that contribute significantly to the predictive ability of the model. In this case we have three significant variables (stayslprec p=.000, getsleprec p=.035, hourwnit p=.007). In this example the major factors influencing whether a person reports having a sleep problem are: difficulty getting to sleep, trouble staying asleep and the number of hours sleep per weeknight. Gender and age did not contribute significantly to the model.

The B values provided in the second column are equivalent to the B values obtained in a multiple regression analysis. These are the values that you would use in an equation to calculate the probability of a case falling into a specific category. You should check whether your B values are positive or negative. This will tell you about the direction of the relationship (which factors increase the likelihood of a yes answer and which factors decrease it). If you have coded all your dependent and independent categorical variables correctly (with 0=no, or lack of the characteristic; 1=yes, or the presence of the characteristic), negative B values indicate that an increase in the independent variable score will result in a decreased probability of the case recording a score of 1 in the dependent variable (indicating the presence of sleep problems in this case). In this example the variable measuring the number of hours slept each weeknight showed a negative B value (−.448). This indicates that the more hours a person sleeps per night the less likely it is that they will report having a sleep problem. For the two other significant categorical variables (trouble getting to sleep, trouble staying asleep), the B values are positive. This suggests that people saying they have difficulty getting to sleep or staying asleep are more likely to answer yes to the question whether they consider they have a sleep problem.

The other useful piece of information in the **Variables in the Equation** table is provided in the Exp(B) column. These values are the odds ratios (OR) for each of your independent variables. According to Tabachnick and Fidell (2001), the odds ratio is 'the increase (or decrease if the ratio is less than one) in odds of being in one outcome category when the value of the predictor increases by one unit' (p. 548). In our example, the odds of a person answering Yes, they have a sleep problem, is 7.27 times higher for someone who reports having problems staying asleep than for a person who does not have difficulty staying asleep, all other factors being equal.

The hours of sleep a person gets is also a significant predictor, according to the Sig. value (p=.007). The odds ratio for this variable, however, is .639, a value less than 1. This indicates that the more sleep a person gets per night, the less likely he/she is to report a sleep problem. For every extra hour of sleep a person gets, the odds of him/her reporting a sleep problem decreases by a factor of .639, all other factors being equal.

Note here that we have a continuous variable as our predictor; therefore we report the increase (or decrease if less than 1) of the odds for each unit increase (in this case a year) in the predictor variable. For categorical predictor variables we are comparing the odds for the two categories. For categorical variables with more than two categories, each category is compared with the reference group (usually the group coded with the lowest value if you have specified First in the Contrast section of the Define Categorical Variables dialogue box).

For each of the odds ratios Exp(B) shown in the **Variables in the Equation** table there is a 95 per cent confidence interval (95.0% CI for EXP(B)) displayed, giving a lower value and an upper value. These will need to be reported in your results. In simple terms this is the range of values that we can be 95 per cent confident encompasses the true value of the odds ratio. Remember, the value specified as the odds ratio is only a point estimate or guess at the true value, based on the sample data. The confidence that we have in this being an accurate representation of the true value (from the entire population) is dependent to a large extent on the size of our sample. Small samples will result in very wide confidence intervals around the estimated odds ratio. Much smaller intervals will occur if we have large samples. In this example the confidence interval for our variable trouble staying asleep (stayslprec OR=7.27) ranges from 3.85 to 13.75. So, although we quote the calculated OR as 7.27, we can be 95 per cent confident that the actual value of OR in the population lies somewhere between 3.85 and 13.75, quite a wide range of values. The confidence interval in this case does not contain the value of 1, therefore this result is statistically significant at p<.05. If the confidence interval had contained the value of 1 the odds ratio would not be statistically significant—we could not rule out the possibility that the true odds ratio was 1, indicating equal probability of the two responses (yes/no).

The last table in the output, labelled **Casewise List**, gives you information about cases in your sample for whom the model does not fit well. Cases with ZResid values above 2 are shown in the table (in this example showing case numbers 42, 224, 227, 235, 265). Cases with values above 2.5 (or less than –2.5) should be examined more closely, as these are clear outliers (given that 99 per cent of cases will have values between –2.5 and +2.5). You can see from the other information in the casewise list that one of the cases (42) was predicted to be in the Yes (I have a sleep problem) category, but in reality (in the **Observed** column) was found to answer the question with a No. The remainder of the outliers were all predicted to answer no, but instead answered Yes. For all these cases (and certainly for ZResid values over 2.5), it would be a good idea to check the information entered and to find out more about them. You may find that there are certain groups of cases for which the model does not work well (e.g. those people on shiftwork). You may need to consider removing cases with very large ZResid values from the data file and repeating the analysis.

Presenting the results from logistic regression

There are many different ways of presenting the results of logistic regression, depending on the type of analysis that you have conducted and purpose of your report. In journal articles you will often find only a brief description of the results, but for a thesis you would need to present a more detailed description. Tabachnick and Fidell (2001) provide a few examples of results presentations (standard or direct logistic regression on p. 562, and sequential logistic regression on p. 573).

As a minimum, you should probably present information on the type of analysis conducted, the overall model fit (chi-square value, degrees of freedom, N value, p value). Sometimes you will also see the –2 Log Likelihood value specified (available from the Model Summary table in SPSS). You should include information about the classification of cases: a table could be included or just describe the overall percentage correct, and the percentage correctly classified for each group. The pseudo R square values (Cox & Snell, Nagelkerke) could be included to provide some information about the percentage of variance explained. All of the details provided in the **Variables in the Equation** table of the SPSS output should be included (B, S.E. Wald, df, Sig., Exp(B), 95% CI). You would also need a verbal description of the results of the predictors (and their odds ratios) to help the reader understand your results. Indicate which predictors were significant, and the direction of their influence (did they increase or decrease the odds?).

References

Hosmer, D. W., & Lemeshow, S. (2000). *Applied logistic regression.* New York: Wiley.

Peat, J. (2001). *Health science research: A handbook of quantitative methods.* Sydney: Allen & Unwin.

Tabachnick, B. G., & Fidell, L. S. (2001). *Multivariate statistics* (4th edn). Boston: Allyn & Bacon. Chapter 12.

Wright, R. E. (1995) Logistic regression. In L. G. Grimm, & P. R. Yarnold (Eds). *Reading and understanding multivariate statistics.* Washington, DC: American Psychological Association. Chapter 7.

15 Factor analysis

Factor analysis is different from many of the other techniques presented in this book. It is not designed to test hypotheses or to tell you whether one group is significantly different from another. It is included in the SPSS package as a 'data reduction' technique. It takes a large set of variables and looks for a way that the data may be 'reduced' or summarised using a smaller set of factors or components. It does this by looking for 'clumps' or groups among the inter-correlations of a set of variables. This is an almost impossible task to do 'by eye' with anything more than a small number of variables.

This family of factor analytic techniques has a number of different uses. It is used extensively by researchers involved in the development and evaluation of tests and scales. The scale developer starts with a large number of individual scale items and questions and, by using factor analytic techniques, they can refine and reduce these items to form a smaller number of coherent subscales. Factor analysis can also be used to reduce a large number of related variables to a more manageable number, prior to using them in other analyses such as multiple regression or multivariate analysis of variance.

There are two main approaches to factor analysis that you will see described in the literature—exploratory and confirmatory. Exploratory factor analysis is often used in the early stages of research to gather information about (explore) the interrelationships among a set of variables. Confirmatory factor analysis, on the other hand, is a more complex and sophisticated set of techniques used later in the research process to test (confirm) specific hypotheses or theories concerning the structure underlying a set of variables.

The term 'factor analysis' encompasses a variety of different, although related techniques. One of the main distinctions is between what is termed principal components analysis (PCA) and factor analysis (FA). These two sets of techniques are similar in many ways and are often used interchangeably by researchers. Both attempt to produce a smaller number of linear combinations of the original variables in a way that captures (or accounts for) most of the variability in the pattern of correlations. They do differ in a number of ways, however. In principal components analysis the original variables are transformed into a smaller set of linear combinations, with all of the variance in the variables being used. In factor analysis, however, factors are estimated using a mathematical model, where only the shared variance is analysed (see Tabachnick & Fidell, 2001, Chapter 13, for more information on this).

Although both approaches (PCA and FA) often produce similar results, books on the topic often differ in terms of which approach they recommend. Stevens (1996, pp. 362–363) admits a preference for principal components analysis and gives a number of reasons for this. He suggests that it is psychometrically sound and simpler mathematically, and it avoids some of the potential problems with 'factor indeterminancy' associated with factor analysis (Stevens, 1996, p. 363). Tabachnick and Fidell (2001, pp. 610–611), in their review of PCA and FA, conclude that: 'If you are interested in a theoretical solution uncontaminated by unique and error variability, FA is your choice. If on the other hand you want an empirical summary of the data set, PCA is the better choice' (p. 611). I have chosen to demonstrate principal components analysis in this chapter. If you would like to explore the other approaches further, see Tabachnick and Fidell (2001).

Note: Although PCA technically yields components, many authors use the term *factor* to refer to the output of both PCA and FA. So don't assume, if you see the term *factor* when you are reading journal articles, that the author has used FA. Factor analysis is used as a general term to refer to the entire family of techniques.

Another potential area of confusion involves the use of the word 'factor', which has different meanings and uses in different types of statistical analyses. In factor analysis it refers to the group or clump of related variables, while in analysis of variance techniques it refers to the independent variable. These are very different things, despite having the same name, so keep the distinction clear in your mind when you are performing the different analyses.

Steps involved in factor analysis

There are three main steps in conducting factor analysis (I am using the term in a general sense to indicate any of this family of techniques, including principal components analysis).

Step 1: Assessment of the suitability of the data for factor analysis

There are two main issues to consider in determining whether a particular data set is suitable for factor analysis: sample size, and the strength of the relationship among the variables (or items). While there is little agreement among authors concerning how large a sample should be, the recommendation generally is: the larger, the better. In small samples the correlation coefficients among the variables are less reliable, tending to vary from sample to sample. Factors obtained from small data sets do not generalise as well as those derived from larger samples. Tabachnick and Fidell (2001) review this issue and suggest that 'it is comforting to have at least 300 cases for factor analysis' (p. 588). However, they do concede that a smaller sample size (e.g. 150 cases) should be sufficient if solutions have several high loading marker variables (above .80). Stevens (1996, p. 372) suggests that the sample size requirements advocated by researchers have been reducing over the years as more research has been done on the topic. He makes a number of recommendations concerning the reliability of factor structures and the sample size requirements (see Stevens, 1996, Chapter 11).

Some authors suggest that it is not the overall sample size that is of concern—rather the ratio of subjects to items. Nunnally (1978) recommends a 10 to 1 ratio: that is, 10 cases for each item to be factor analysed. Others suggest that 5 cases for each item is adequate in most cases (see discussion in Tabachnick & Fidell, 2001). I would recommend that you do more reading on the topic, particularly if you have a small sample (smaller than 150), or lots of variables.

The second issue to be addressed concerns the strength of the inter-correlations among the items. Tabachnick and Fidell recommend an inspection of the correlation matrix for evidence of coefficients greater than .3. If few correlations above this level are found, then factor analysis may not be appropriate. Two statistical measures are also generated by SPSS to help assess the factorability of the data: Bartlett's test of sphericity (Bartlett, 1954), and the Kaiser-Meyer-Olkin (KMO) measure of sampling adequacy (Kaiser, 1970, 1974). The Bartlett's test of sphericity should be significant ($p<.05$) for the factor analysis to be considered appropriate. The KMO index ranges from 0 to 1, with .6 suggested as the minimum value for a good factor analysis (Tabachnick & Fidell, 2001).

Step 2: Factor extraction

Factor extraction involves determining the smallest number of factors that can be used to best represent the interrelations among the set of variables. There are a variety of approaches that can be used to identify (extract) the number of underlying factors or dimensions. Some of the most commonly available extraction techniques (this always conjures up the image for me of a dentist pulling teeth!) are:

- principal components;
- principal factors;
- image factoring;
- maximum likelihood factoring;
- alpha factoring;
- unweighted least squares; and
- generalised least squares.

The most commonly used approach (as noted earlier) is principal components analysis. This will be demonstrated in the example given later in this chapter.

It is up to the researcher to determine the number of factors that he/she considers best describes the underlying relationship among the variables. This involves balancing two conflicting needs: the need to find a simple solution with as few factors as possible; and the need to explain as much of the variance in the original data set as possible. Tabachnick and Fidell (2001) recommend that researchers adopt an exploratory approach, experimenting with different numbers of factors until a satisfactory solution is found.

There are a number of techniques that can be used to assist in the decision concerning the number of factors to retain:

- Kaiser's criterion;
- scree test; and
- parallel analysis.

Kaiser's criterion

One of the most commonly used techniques is known as Kaiser's criterion, or the eigenvalue rule. Using this rule, only factors with an eigenvalue of 1.0 or more are retained for further investigation (this will become clearer when you see the example presented in this chapter). The eigenvalue of a factor represents the amount of the total variance explained by that factor. Kaiser's criterion has been criticised, however, as resulting in the retention of too many factors in some situations.

Scree test

Another approach that can be used is Catell's scree test (Catell, 1966). This involves plotting each of the eigenvalues of the factors (SPSS can do this for you) and inspecting the plot to find a point at which the shape of the curve changes direction and becomes horizontal. Catell recommends retaining all factors above the elbow, or break in the plot, as these factors contribute the most to the explanation of the variance in the data set.

Parallel analysis

An additional technique gaining popularity, particularly in the social science literature (e.g. Choi, Fuqua, & Griffin, 2001; Stober, 1998), is Horn's parallel analysis (Horn, 1965). Parallel analysis involves comparing the size of the eigenvalues with those obtained from a randomly generated data set of the same size. Only those eigenvalues that exceed the corresponding values from the random data set are retained. This approach to identifying the correct number of components to retain has been shown to be the most accurate, with both Kaiser's criterion and Catell's scree test tending to overestimate the number of components (Hubbard & Allen, 1987; Zwick & Velicer, 1986). If you intend to publish your results in a journal article in the psychology or educational fields you will need to use, and report, the results of parallel analysis. Many journals (e.g. *Educational and Psychological Measurement, Journal of Personality Assessment*) are now making it a requirement before they will consider a manuscript for publication.

The use of these three techniques is demonstrated in the worked example presented later in this chapter.

Step 3: Factor rotation and interpretation

Once the number of factors have been determined, the next step is to try to interpret them. To assist in this process the factors are 'rotated'. This does not change the underlying solution—rather, it presents the pattern of loadings in a manner that is easier to interpret. Unfortunately, SPSS does not label or interpret each of the factors for you. It just shows you which variables 'clump together'. From your understanding of the content of the variables (and underlying theory and past research), it is up to you to propose possible interpretations.

There are two main approaches to rotation, resulting in either orthogonal (uncorrelated) or oblique (correlated) factor solutions. According to Tabachnick and Fidell (2001), orthogonal rotation results in solutions that are easier to interpret and to report; however, they do require the researcher to assume that the underlying constructs are independent (not correlated). Oblique approaches allow for the factors to be correlated, but they are more difficult to interpret, describe and report (Tabachnick & Fidell, 2001, p. 618). In practice, the two approaches (orthogonal and oblique) often result in very similar solutions, particularly when the pattern of correlations among the items is clear (Tabachnick & Fidell, 2001). Many researchers conduct both orthogonal and oblique rotations and then report the clearest and easiest to interpret. You are hoping for what Thurstone (1947) refers to as 'simple structure'. This involves each of the variables loading strongly on only one component, and each component being represented by a number of strongly loading variables.

Within the two broad categories of rotational approaches there are a number of different rotational techniques provided by SPSS (orthogonal: Varimax, Quartimax, Equamax; oblique: Direct Oblimin, Promax). The most commonly used orthogonal approach is the Varimax method, which attempts to minimise the number of variables that have high loadings on each factor. The most commonly used oblique technique is Direct Oblimin. For a comparison of the characteristics of each of these approaches, see Tabachnick and Fidell (2001, p. 615). In the example presented in this chapter both Varimax rotation and Oblimin rotation will be demonstrated.

Additional resources

In this chapter only a very brief overview of factor analysis is provided. Although I have attempted to simplify it here, factor analysis is actually a sophisticated and complex family of techniques. If you are intending to use factor analysis with your own data I strongly recommend that you read up on the technique in more depth. See the References at the end of this chapter.

Details of example

To demonstrate the use of factor analysis I will explore the underlying structure of one of the scales included in the survey.sav data file provided on the website accompanying this book (see p. xi). The survey was designed to explore the factors that affect respondents' psychological adjustment and wellbeing (see the Appendix for a full description of the study). One of the scales that was used was the Positive and Negative Affect scale (PANAS: Watson, Clark, & Tellegen, 1988) (see Figure 15.1). This scale consists of twenty adjectives describing different mood states, ten positive (proud, active, determined) and ten negative (nervous, irritable, upset). The authors of the scale suggest that the PANAS consists of two underlying dimensions (or factors): positive affect and negative affect. To explore this structure with the current community sample, the items of the scale will be subjected to principal components analysis (PCA), a form of factor analysis that is commonly used by researchers interested in scale development and evaluation.

If you wish to follow along with the steps described in this chapter, you should start SPSS and open the file labelled survey.sav on the website that accompanies this book (see p. xi; this file can be opened only in SPSS). The variables that are used in this analysis are labelled: pn1 to pn20. The scale used in the survey is presented in Figure 15.1. You will need to refer to these individual items when attempting to interpret the factors obtained. For full details and references for the scale, see the Appendix.

Figure 15.1

Positive and Negative Affect scale (PANAS)

This scale consists of a number of words that describe different feelings and emotions. For each item indicate to what extent you have felt this way during the past few weeks. Write a number from 1 to 5 on the line next to each item.

very slightly or not at all	a little	moderately	quite a bit	extremely
1	2	3	4	5

1. ____ interested
2. ____ upset
3. ____ scared
4. ____ proud
5. ____ ashamed
6. ____ determined
7. ____ active

8. ____ distressed
9. ____ strong
10. ____ hostile
11. ____ irritable
12. ____ inspired
13. ____ attentive
14. ____ afraid

15. ____ excited
16. ____ guilty
17. ____ enthusiastic
18. ____ alert
19. ____ nervous
20. ____ jittery

Summary for factor analysis

Example of research question: What is the underlying factor structure of the Positive and Negative Affect scale? Past research suggests a two-factor structure (positive affect/negative affect). Is the structure of the scale in this study, using a community sample, consistent with this previous research?

What you need: A set of correlated continuous variables.

What it does: Factor analysis attempts to identify a small set of factors that represents the underlying relationships among a group of related variables.

Assumptions:

1. *Sample size*. Ideally the overall sample size should be 150+ and there should be a ratio of at least five cases for each of the variables (see discussion in Step 1 earlier in this chapter).

2. *Factorability of the correlation matrix*. To be considered suitable for factor analysis the correlation matrix should show at least some correlations of r=.3 or greater. The Bartlett's test of sphericity should be statistically significant at p<.05 and the Kaiser-Meyer-Olkin value should be .6 or above. These values are presented as part of the output from factor analysis.

3. *Linearity*. Because factor analysis is based on correlation, it is assumed that the relationship between the variables is linear. It is certainly not practical to check scatterplots of all variables with all other variables. Tabachnick and Fidell (2001) suggest a 'spot check' of some combination of variables. Unless there is clear evidence of a curvilinear relationship, you are probably safe to proceed, providing you have an adequate sample size and ratio of cases to variables (see Assumption 1).

4. *Outliers among cases*. Factor analysis can be sensitive to outliers, so as part of your initial data screening process (see Chapter 6) you should check for these and either remove or recode to a less extreme value.

Procedure for factor analysis

As discussed previously, factor analysis involves a number of steps: assessment of the data, factor extraction and factor rotation. To cover the SPSS procedures required for these steps, this section is divided into two parts. In Part 1 the procedures involved in the assessment of the data and the extraction of the factors will be presented and the output discussed. In Part 2 the additional procedures required to rotate and to interpret the factors will be covered.

Part 1: Assessing the data and extracting the factors

The first step when performing a factor analysis is to assess the suitability of the data for factor analysis. This involves inspecting the correlation matrix for coefficients of .3 and above, and calculating the Kaiser-Meyer-Olkin Measure of Sampling Adequacy (KMO) and Barlett's Test of Sphericity. This information can be obtained from SPSS in the same analysis as used for Factor Extraction. The second step involves determining how many underlying factors there are in the set of variables.

Procedure (Part 1)

1. From the menu at the top of the screen click on: **Analyze**, then click on **Data Reduction**, then on **Factor**.

2. Select all the required variables (or items on the scale). In this case I would select the items that make up the PANAS scale (pn1 to pn20). Move them into the **Variables** box.

3. Click on the **Descriptives** button.
   - In the section marked **Correlation Matrix**, select the options **Coefficients** and **KMO and Bartlett's test of sphericity** (these will test some of your assumptions).
   - In the **Statistics** section, make sure that **Initial Solution** is ticked. Click on **Continue**.

4. Click on the **Extraction** button.
   - In the **Method** section make sure **Principal components** is listed.
   - In the **Analyze** section make sure the **Correlation matrix** option is selected.
   - In the **Display** section, click on **Screeplot** and make sure the **Unrotated factor solution** option is also selected.
   - In the **Extract** section make sure the **Eigenvalues over 1** button is selected. Click on **Continue**.

5. Click on the **Options** button.
   - In the **Missing Values** section click on **Exclude cases pairwise**.
   - In the **Coefficient Display Format** section click on **Sorted by size** and **Suppress absolute values less than___**. Type the value of .3 in the box. This means that only loadings above .3 will be displayed, making the output easier to interpret.

6. Click on **Continue** and then **OK**.

The output generated from this procedure is shown below. Only selected output is displayed.

Correlation Matrix

		PN1	PN2	PN3	PN4	PN5	PN6	PN7	PN8	PN9	PN10
Correl	PN1	1.000	-.139	-.152	.346	-.071	.352	.407	-.250	.416	-.122
	PN2	-.139	1.000	.462	-.141	.271	-.127	-.197	.645	-.188	.411
	PN3	-.152	.462	1.000	-.102	.247	-.097	-.255	.494	-.200	.234
	PN4	.346	-.141	-.102	1.000	-.156	.295	.331	-.152	.396	-.056
	PN5	-.071	.271	.247	-.156	1.000	-.067	-.248	.278	-.201	.258
	PN6	.352	-.127	-.097	.295	-.067	1.000	.329	-.048	.426	.077
	PN7	.407	-.197	-.255	.331	-.248	.329	1.000	-.232	.481	-.093
	PN8	-.250	.645	.494	-.152	.278	-.048	-.232	1.000	-.181	.380
	PN9	.416	-.188	-.200	.396	-.201	.426	.481	-.181	1.00	-.070
	PN10	-.122	.411	.234	-.056	.258	.077	-.093	.380	-.070	1.000
	PN11	-.210	.500	.333	-.179	.266	-.043	-.214	.464	-.210	.583
	PN12	.482	-.204	-.135	.315	-.063	.401	.400	-.175	.407	-.074
	PN13	.491	-.171	-.165	.329	-.137	.336	.391	-.199	.427	-.114
	PN14	-.151	.406	.810	-.107	.302	-.090	-.271	.459	-.198	.263
	PN15	.413	-.136	-.085	.317	-.062	.276	.329	-.098	.362	-.067
	PN16	-.177	.314	.330	-.121	.539	-.099	-.221	.378	-.164	.314
	PN17	.562	-.208	-.190	.368	-.156	.396	.484	-.218	.465	-.134
	PN18	.466	-.196	-.181	.338	-.189	.451	.458	-.234	.462	-.066
	PN19	-.148	.459	.560	-.124	.285	-.050	-.234	.480	-.198	.339
	PN20	-.176	.425	.424	-.171	.245	-.025	-.204	.431	-.219	.367

KMO and Bartlett's Test

Kaiser-Meyer-Olkin Measure of Sampling Adequacy.		.874
Bartlett's Test of Sphericity	Approx. Chi-Square	3966.539
	df	190
	Sig.	.000

Total Variance Explained

Component	Initial Eigenvalues			Extraction Sums of Squared Loadings		
	Total	% of Variance	Cumulative %	Total	% of Variance	Cumulative %
1	6.250	31.249	31.249	6.250	31.249	31.249
2	3.396	16.979	48.228	3.396	16.979	48.228
3	1.223	6.113	54.341	1.223	6.113	54.341
4	1.158	5.788	60.130	1.158	5.788	60.130
5	.898	4.490	64.619			
6	.785	3.926	68.546			
7	.731	3.655	72.201			
8	.655	3.275	75.476			
9	.650	3.248	78.724			
10	.601	3.004	81.728			
11	.586	2.928	84.656			
12	.499	2.495	87.151			
13	.491	2.456	89.607			
14	.393	1.964	91.571			
15	.375	1.875	93.446			
16	.331	1.653	95.100			
17	.299	1.496	96.595			
18	.283	1.414	98.010			
19	.223	1.117	99.126			
20	.175	.874	100.000			

Extraction Method: Principal Component Analysis.

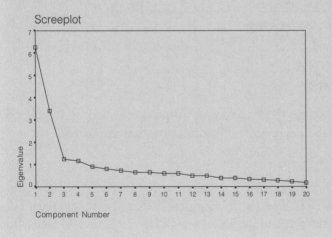

Screeplot

Component Matrix[a]

	Component			
	1	2	3	4
PN17	.679	.474		
PN18	.639	.404		
PN7	.621			
PN8	-.614	.420		
PN9	.609	.323		
PN13	.607	.413		
PN1	.600	.381		
PN2	-.591	.408		
PN3	-.584	.449	-.457	
PN14	-.583	.456	-.451	
PN12	.582	.497		
PN19	-.569	.545		
PN11	-.554	.366	.462	
PN20	-.545	.459		
PN4	.474			
PN15	.477	.483		
PN6	.432	.437		
PN10	-.416	.426	.563	
PN5	-.429			.649
PN16	-.474	.357		.566

Extraction Method: Principal Component Analysis.

Interpretation of output (Part 1)

As with most SPSS procedures there is a lot of output generated. In this section I will take you through the key pieces of information that you need.

Step 1

In the **Correlation Matrix** table, look for correlation coefficients of .3 and above (see Assumption 2). If you don't find any in your matrix then you should reconsider the use of factor analysis. You should also check that the **Kaiser-Meyer-Olkin Measure of Sampling Adequacy** (KMO) value is .6 or above. The **Barlett's Test of Sphericity** value should be significant (i.e. the Sig. value should be .05 or smaller). In this example the KMO value is .874, and the Bartlett's test is significant (p=.000), therefore factor analysis is appropriate.

Step 2

To determine how many components (factors) to 'extract' we need to consider a few pieces of information provided in the output. Using Kaiser's criterion (explained earlier in this chapter), we are interested only in components that have an eigenvalue of 1 or more. To determine how many components meet this criterion we need to look in the Total Variance Explained table. Scan down the values provided in the first set of columns, labelled Initial Eigenvalues. The eigenvalues for each component are listed. In this example only the first four components recorded eigenvalues above 1 (6.25, 3.396, 1.223, 1.158). These four components explain a total of 60.13 per cent of the variance (see Cumulative % column).

Step 3

Often, using the Kaiser criterion, you will find that too many components are extracted, so it is important to also look at the screeplot provided by SPSS. As you will recall, what you need to look for is a change (or elbow) in the shape of the plot. Only components above this point are retained. In this example there is quite a clear break between the second and third components. Components 1 and 2 explain or capture much more of the variance than the remaining components. From this plot, I would recommend retaining (extracting) only two components. There is also another little break after the fourth component. Depending on the research context, this might also be worth exploring. Remember, factor analysis is used as a data exploration technique, SO the interpretation and the use you put it to is up to your judgment, rather than any hard and fast statistical rules.

Step 4

The third way of determining the number of factors to retain is parallel analysis (see discussion earlier in this chapter). For this procedure you need to use the list of eigenvalues provided in the Total Variance Explained table, and some additional information that you must get from another little statistical program (developed by Marley Watkins, 2000) that is available from the website for this book (*www.allenandunwin.com/spss.htm*). Follow the links to the Additional Material site and download the file **MonteCarlo%20PA.exe.zip** onto your computer. Unzip this onto your hard drive and click on the file MonteCarloPA.exe. A program will start that is called Monte Carlo PCA for Parallel Analysis.

You will be asked for three pieces of information: the number of variables you are analysing (in this case 20); the number of subjects in your sample (in this case 435); and the number of replications (specify 100). Click on **Calculate**. Behind the scenes this program will generate 100 sets of random data of the same size as your real data file (20 variables × 435 cases). It will calculate the average eigenvalues for these 100 randomly generated samples and print these out for you. See printout below.

Table X: Output from parallel analysis

7/03/2004   11:58:37 AM
Number of variables: 20
Number of subjects: 435
Number of replications: 100

Eigenvalue	Random Eigenvalue	Standard Dev
1	1.3984	.0422
2	1.3277	.0282
3	1.2733	.0262
4	1.2233	.0236
5	1.1832	.0191
6	1.1433	.0206
7	1.1057	.0192
8	1.0679	.0193
9	1.0389	.0186
10	1.0033	.0153
11	0.9712	.0180
12	0.9380	.0175
13	0.9051	.0187
14	0.8733	.0179
15	0.8435	.0187
16	0.8107	.0185
17	0.7804	.0190
18	0.7449	.0194
19	0.7090	.0224
20	0.6587	.0242

7/03/2004   11:58:50 AM
MonteCarlo PCA for Parallel Analysis
Watkins, M. W. (2000). MonteCarlo PCA for parallel analysis [computer software]. State College, PA: Ed & Psych Associates.

Your job is to systematically compare the first eigenvalue you obtained in SPSS with the corresponding first value from the random results generated by parallel analysis. If your value is larger than the criterion value from parallel analysis, then you retain this factor; if it is less, then you reject it. The results for this example can be summarised in Table 15.1.

Table 15.1

Comparison of eigenvalues from principal components analysis (PCA) and the corresponding criterion values obtained from parallel analysis

Component number	Actual eigenvalue from PCA	Criterion value from parallel analysis	Decision
1	6.250	1.3984	accept
2	3.396	1.3277	accept
3	1.223	1.2733	reject
4	1.158	1.2233	reject
5	.898	1.1832	reject

The results of parallel analysis support our decision from the screeplot to retain only two factors for further investigation.

Step 5

Moving back to our SPSS output, the final table we need to look at is the **Component Matrix**. This shows the loadings of each of the items on the four components. SPSS uses the Kaiser crtierion (retain all components with eigenvalues above 1) as the default. You will see from this table that most of the items load quite strongly (above .4) on the first two components. Very few items load on Components 3 and 4. This supports our conclusion from the screeplot to retain only two factors for further investigation.

In the next section the procedure for rotating these two factors (to aid interpretation) will be demonstrated.

Part 2: Factor rotation and interpretation

Once the number of factors have been determined, the next step is to try to interpret them. To assist in this process the factors are 'rotated'. This does not change the underlying solution—rather, it presents the pattern of loadings in a manner that is easier to interpret. In this example two components will be extracted and rotated. There are a number of different rotation techniques (explained earlier in this chapter). Here I will use Varimax rotation, first as the output is simpler to interpret. I will then repeat the analysis using Oblimin rotation. This is an oblique rotational technique (which does not assume that the factors are uncorrelated). The output from Oblimin rotation will allow us to determine how strongly inter-correlated the factors actually are, and therefore which of the rotational techniques is more appropriate.

Procedure (Part 2)

The procedure described here assumes you have just completed the procedures described in Step 1 and 2 outlined earlier in this chapter. All of the variables should still be listed and the options that you selected last time should still be in place.

1. From the menu at the top of the screen click on: **Analyze**, then click on **Data Reduction**, then on **Factor**.

2. Check that all the required variables (or items on the scale) are still listed in the **Variables** box (pn1 to pn20).

3. Click on the **Descriptives** button.
   - To save repeating the same analyses as obtained in the previous SPSS output you should remove the tick in the **Initial Solution** box, the **Coefficients** box and the **KMO and Bartlett's Test** box. To do this just click on the box with the tick and it should disappear.
   - Click on **Continue**.

Strange looking numbers

In your output you might come across strange looking numbers which take the form: 1.24E−02. These numbers have been written in scientific notation. They can easily be converted to a 'normal' number for presentation in your results. The number after the E tells you how many places you need to shift the decimal point. If the number after the E is negative, shift the decimal point to the left; if it is positive, shift it to the right. Therefore, 1.24E−02 becomes 0.124. The number 1.24E02 would be written as 124.

You can prevent these small numbers from being displayed in this strange way. Go to **Edit, Options** and put a tick in the box labelled: **No scientific notation for small numbers in tables**. Rerun the analysis and normal numbers should appear in all your output.

4. Click on the **Extraction** button.
   - In the **Method** section make sure **Principal Components** is listed.
   - In the **Analyze** section make sure the **Correlation matrix** option is selected.
   - In the **Display** section, remove the tick from the **Screeplot** and the **Unrotated factor solution**.
   - In the **Extract** section select the **Number of Factors** option. In the box type in the number of factors that you wish to extract (in this case 2).
   - Click on **Continue**.

5. Click on the **Options** button.
   - In the Missing Values section click on **Exclude cases pairwise**.
   - In the **Coefficient Display Format** section make sure that there is a tick in **Sorted by size** and **Suppress absolute values less than .3**.
   - Click on **Continue**.

6. Click on the **Rotation** button.
   - In the **Method** section click on **Varimax**.

7. Click on **Continue** and then **OK**.

The output generated is shown below.

Total Variance Explained

Component	Initial Eigenvalues			Rotation Sums of Squared Loadings		
	Total	% of Variance	Cumulative %	Total	% of Variance	Cumulative %
1	6.250	31.249	31.249	4.886	24.429	24.429
2	3.396	16.979	48.228	4.760	23.799	48.228
3	1.223	6.113	54.341			
4	1.158	5.788	60.130			
5	.898	4.490	64.619			
6	.785	3.926	68.546			
7	.731	3.655	72.201			
8	.655	3.275	75.476			
9	.650	3.248	78.724			
10	.601	3.004	81.728			
11	.586	2.928	84.656			
12	.499	2.495	87.151			
13	.491	2.456	89.607			
14	.393	1.964	91.571			
15	.375	1.875	93.446			
16	.331	1.653	95.100			
17	.299	1.496	96.595			
18	.283	1.414	98.010			
19	.223	1.117	99.126			
20	.175	.874	100.000			

Extraction Method: Principal Component Analysis.

Rotated Component Matrix [a]

	Component	
	1	2
PN17	.819	
PN12	.764	
PN18	.741	
PN13	.724	
PN1	.697	
PN15	.679	
PN9	.663	
PN7	.617	
PN6	.614	
PN4	.541	
PN19		.787
PN14		.732
PN3		.728
PN8		.728
PN20		.708
PN2		.704
PN11		.647
PN10		.595
PN16		.585
PN5		.493

Extraction Method: Principal Component Analysis.

Rotation Method: Varimax with Kaiser Normalization.

[a.] Rotation converged in 3 iterations.

Interpretation of output (Varimax rotation)

- In the table labelled **Total Variance Explained** you will notice that there are now only two components listed in the right-hand section (as compared with four, in the previous unrotated output). This is because we asked SPSS to select only two components for rotation. You will see that the distribution of the variance explained has also been adjusted after rotation. Component 1 now explains 24.43 per cent of the variance and Component 2 explains 23.80 per cent. The total variance explained (48.2 per cent) does not change after rotation, just the way that it is distributed between the two components.
- In the **Rotated Component Matrix** you will see the loadings of each of the variables on the two factors that were selected. Look for the highest loading variables on each of the component—these can be used to help you identify the nature of the underlying latent variable represented by each component.

In this example the main loadings on Component 1 are items 17, 12, 18 and 13. If you refer back to the actual items themselves (presented earlier in this chapter) you will see that these are all positive affect items (enthusiastic, inspired, alert, attentive). The main items on Component 2 (19, 14, 3, 8, 20) are negative affect items (nervous, afraid, scared, jittery). In this case identification and labelling of the two components is easy. This is not always the case, however.

Watch out for situations where you have lots of cross loadings (variables that load on more than one component)—you may need to go back and play around with the number of components that you have decided to extract (try both one more and one less).

Oblimin rotation

To obtain the results of Oblimin rotation (which allows your components to be correlated with one another) you should follow the instructions given earlier, but when you click on the **Rotation** button in the Method section choose **Direct Oblimin**.

Selected parts of the output are shown below.

Pattern Matrix[a]

	Component	
	1	2
pn17	.825	
pn12	.781	
pn18	.742	
pn13	.728	
pn15	.703	
pn1	.698	
pn9	.656	
pn6	.635	
pn7	.599	
pn4	.540	
pn19		.806
pn14		.739
pn3		.734
pn8		.728
pn20		.718
pn2		.704
pn11		.645
pn10		.613
pn16		.589
pn5		.490

Extraction Method: Principal Component Analysis.
Rotation Method: Oblimin with Kaiser Normalization.
a. Rotation converged in 6 iterations.

Structure Matrix

	Component	
	1	2
pn17	.828	
pn12	.763	
pn18	.755	
pn13	.733	
pn1	.710	
pn9	.683	
pn15	.670	
pn7	.646	-.338
pn6	.605	
pn4	.553	
pn19		.784
pn8		.742
pn14		.740
pn3		.737
pn2		.717
pn20		.712
pn11		.661
pn16		.593
pn10		.590
pn5		.505

Extraction Method: Principal Component Analysis.
Rotation Method: Oblimin with Kaiser Normalization.

Component Correlation Matrix

Component	1	2
1	1.000	-.277
2	-.277	1.000

Extraction Method: Principal Component Analysis.
Rotation Method: Oblimin with Kaiser Normalization.

Interpretation of output (Oblimin rotation)

You will notice that the output from Oblimin rotation is different from that produced by Varimax rotation. There are three main tables you need to consider: Pattern Matrix, Structure Matrix and Component Correlation Matrix.

First have a look at the **Component Correlation Matrix** (at the end of the output). This shows you the strength of the relationship between the two factors (in this case the value is quite low, at –.277). This gives us information to decide whether it was reasonable to assume that the two components were not related (the assumption underlying the use of the previous Varimax rotation) or whether it is necessary to use, and report, the more complex Oblimin rotation solution shown here. In this case the correlation between the two components is quite low, so we would expect very similar solutions from the Varimax and Oblimin rotation. If, however, your components are more strongly correlated (e.g. above .3) then you may find discrepancies between the results of the two approaches to rotation. If that is the case, you need to report the Oblimin rotation.

Let's have a look at the remaining output from Oblimin rotation. You should have two tables. The first table, labelled **Pattern Matrix,** is similar to that provided in the previous Varimax rotated solution. It shows the factor loadings of each of the variables and can be interpreted in much the same way. Look for the highest loading items on each component to identify and label the component. The **Structure Matrix** table, which is unique to the Oblimin output, provides information about the correlation between variables and factors. If you need to present the Oblimin rotated solution in your output you must present both of these tables.

Warning

The output in this example is a very 'clean' result. Each of the variables loaded strongly on only one component, and each component was represented by a number of strongly loading variables (an example of 'simple structure'). For a discussion of this topic, see Tabachnick and Fidell (2001, p. 622). Unfortunately, with your own data you will not always have such a straightforward result. Often you will find that variables load moderately on a number of different components, and some components will have only one or two variables loading on them. In cases such as this you may need to consider rotating a different number of components (e.g. one more and one less) to see whether a more optimal solution can be found. If you find that some variables just do not load on the components obtained, you may also need to consider removing them, and repeating the analysis. You should read as much as you can on the topic to help you make these decisions.

Presenting the results from factor analysis

The information you provide in your results section is dependent on your discipline area, the type of report you are preparing and where it will be presented. If you are publishing in the areas of psychology and education particularly, there are quite strict requirements for what needs to be included in a journal article that involves the use of factor analysis. Henson and Roberts (in press) provide a detailed summary of what should be reported. This includes details of the method of factor extraction used, the criteria used to determine the number of factors (this should include parallel analysis), the type of rotation technique used (e.g. Varimax, Oblimin), the total variance explained, the initial eigenvalues, and the eigenvalues after rotation. A table of loadings should be included showing all values (not just those above .3). For the Varimax rotated solution the table should be labelled: 'pattern/structure coefficients'. If Oblimin rotation was used, then both the Pattern Matrix and the Structure matrix should be presented in full, along with information on the correlations among the factors.

The results of the output obtained in the example above could be presented as follows:

The 20 items of the Positive and Negative Affect scale (PANAS) were subjected to principal components analysis (PCA) using SPSS Version 12. Prior to performing PCA the suitability of data for factor analysis was assessed. Inspection of the correlation matrix revealed the presence of many coefficients of .3 and above. The Kaiser-Meyer-Oklin value was .87, exceeding the recommended value of .6 (Kaiser, 1970, 1974) and the Barlett's Test of Sphericity (Bartlett, 1954) reached statistical significance, supporting the factorability of the correlation matrix.

Principal components analysis revealed the presence of four components with eigenvalues exceeding 1, explaining 31.2 per cent, 17 per cent, 6.1 per cent and 5.8 per cent of the variance respectively. An inspection of the screeplot revealed a clear break after the second component. Using Catell's (1966) scree test, it was decided to retain two components for further investigation. This was further supported by the results of Parallel Analysis, which showed only two components with eigenvalues exceeding the corresponding criterion values for a randomly generated data matrix of the same size (20 variables × 435 respondents).

TABLE Pattern/structure for coefficients

Varimax Rotation of Two Factor Solution for PANAS Items

Item	Component 1 Positive Affect	Component 2 Negative Affect
Enthusiastic	.819	−.127
Inspired	.764	−.043
Alert	.741	−.150
Attentive	.724	−.121
Interested	.697	−.140
Excited	.679	.019
Strong	.663	−.187
Active	.617	−.254
Determined	.614	.017
Proud	.541	−.120
Nervous	−.034	.787
Afraid	−.106	.732
Scared	−.112	.728
Distressed	−.153	.728
Jittery	−.077	.708
Upset	−.145	.704
Irritable	−.147	.647
Hostile	−.006	.595
Guilty	−.095	.585
Ashamed	−.123	.493
% of variance explained	24.43%	23.80%

To aid in the interpretation of these two components, Varimax rotation was performed. The rotated solution revealed the presence of simple structure (Thurstone, 1947), with both components showing a number of strong loadings and all variables loading substantially on only one component. The two-component solution explained a total of 48.2 per cent of the variance, with Component 1 contributing 24.43 per cent and Component 2 contributing 23.8 per cent. The interpretation of the two components was consistent with previous research on the PANAS scale, with positive affect items loading strongly on Component 1 and negative affect items loading strongly on Component 2. The results of this analysis support the use of the positive affect items and the negative affect items as separate scales, as suggested by the scale authors (Watson, Clark, & Tellegen, 1988).

You will need to include two additional tables: Table 15.1: Comparison of eigenvalues from principal components analysis (PCA) and the corresponding criterion values obtained from parallel analysis, as presented earlier in this chapter; and the table of loadings for all variables. Before you can do this you will need to rerun the analysis that you chose as your final solution (e.g. either Varimax or Oblimin rotation), but this time you will need to turn off the option to display only coefficients above .3 (see procedures section). Click on **Options,** and in the **Coefficient Display Format** section remove the tick from the second box **Suppress absolute values less than .3.**

If you are presenting the results of this analysis in your thesis (rather than a journal article), you may also need to provide the screeplot and the table of unrotated loadings in the appendix. This would allow the reader of your thesis to see if they agree with your decision to retain only two components.

Additional exercises

Business

Data file: *staffsurvey.sav.* See Appendix for details of the data file.

1. Follow the instructions throughout the chapter to conduct a principal components analysis with Oblimin rotation on the 10 agreement items that make up the Staff Satisfaction Survey (Q1a to Q10a). You will see that, although two factors record eigenvalues over 1, the screeplot indicates that only one component should be retained.

   Run Parallel Analysis using 523 as the number of cases, and 10 as the number of items. The results indicate only one component has an eigenvalue that exceeds the equivalent value obtained from a random data set. This suggests that the items of the Staff Satisfaction Scale are assessing only one underlying dimension (factor).

Health

Data file: *sleep.sav.* See Appendix for details of the data file.

1. Use the procedures shown in Chapter 15 to explore the structure underlying the set of questions designed to assess the impact of sleep problems on various aspects of people's

lives. These items are labelled *impact*1 to *impact*7. Run Parallel Analysis (using 121 as the number of cases, and 7 as the number of items) to check how many factors should be retained.

References

Bartlett, M. S. (1954). A note on the multiplying factors for various chi square approximations. *Journal of the Royal Statistical Society, 16* (Series B), 296–298.

Catell, R. B. (1966). The scree test for number of factors. *Multivariate Behavioral Research, 1*, 245–276.

Choi, N., Fuqua, D., & N. Griffin, B. W. (2001). Exploratory analysis of the structure of scores from the multidimensional scales of perceived self efficacy. *Educational and Psychological Measurement, 61*, 475–489.

Gorsuch, R. L. (1983). *Factor analysis*. Hillsdale, NJ: Erlbaum.

Henson, R. K., & Roberts, J. K. (in press). Exploratory factor analysis reporting practices in published research. In B. Thompson (Ed.), *Advances in social science methodology* (Vol. 6). Stamford, CT: JAI Press.

Horn, J. L. (1965). A rationale and test for the number of factors in factor analysis. *Psychometrika, 30*, 179–185.

Hubbard, R., & Allen, S. J. (1987). An empirical comparison of alternative methods for principal component extraction. *Journal of Business Research, 15*, 173–190.

Kaiser, H. (1970). A second generation Little Jiffy. *Psychometrika, 35*, 401–415.

Kaiser, H. (1974). An index of factorial simplicity. *Psychometrika, 39*, 31–36.

Nunnally, J. O. (1978). *Psychometric Theory*. New York: McGraw-Hill.

Stevens, J. (1996). *Applied multivariate statistics for the social sciences* (3rd edn). Mahwah, NJ: Lawrence Erlbaum. Chapter 11.

Stober, J. (1998). The Frost multidimensional perfectionism scale revisited: more perfect with four (instead of six) dimensions. *Personality and Individual Differences, 24*, 481–491.

Tabachnick, B. G., & Fidell, L. S. (2001). *Using multivariate statistics* (4th edn). New York: HarperCollins. Chapter 13.

Thurstone, L. L. (1947). *Multiple factor analysis*. Chicago: University of Chicago Press.

Watkins, M. W. (2000). *Monte carlo PCA for parallel analysis* [computer software]. State College, PA: Ed & Psych Associates.

Watson, D., Clark, L. A., & Tellegen, A. (1988). Development and validation of brief measures of positive and negative affect: the PANAS scales. *Journal of Personality and Social Psychology, 54*, 1063–1070.

Zwick, W. R., & Velicer, W. F. (1986). Comparison of five rules for determining the number of components to retain. *Psychological Bulletin, 99*, 432–442.

Part Five

Statistical techniques to compare groups

In Part Five of this book we will be exploring some of the techniques available in SPSS to assess differences between groups or conditions. The techniques used are quite complex, drawing on a lot of underlying theory and statistical principles. Before you start your analysis using SPSS it is important that you have at least a basic understanding of the statistical techniques that you intend to use. There are many good statistical texts available that can help you with this (a list of some suitable books is provided in the References for this part and in the Recommended references at the end of the book). It would be a good idea to review this material now. This will help you understand what SPSS is calculating for you, what it means and how to interpret the complex array of numbers generated in the output. In the remainder of this part and the following chapters I have assumed that you have a basic grounding in statistics and are familiar with the terminology used.

Techniques covered in Part Five

There is a whole family of techniques that can be used to test for significant differences between groups. Although there are many different statistical techniques available in the SPSS package, only the main techniques are covered here:

- Independent-samples t-test;
- Paired-samples t-test;
- One-way analysis of variance (between groups);
- one-way analysis of variance (repeated measures);
- two-way analysis of variance (between groups);
- mixed between-within groups analysis of variance;
- multivariate analysis of variance (MANOVA);
- one-way, and two-way analysis of covariance (ANCOVA); and
- non-parametric techniques.

In Chapter 10 you were guided through the process of deciding which statistical technique would suit your research question. This depends on the nature of your research question, the type of data you have and the number of variables and groups you have. (If you have not read through that chapter, then you should do so before proceeding any further.)

Some of the key points to remember when choosing which technique is the right one for you are as follows:

- T-tests are used when you have only *two* groups (e.g. males/females).
- Analysis of variance techniques are used when you have *two or more* groups.
- Paired-samples or repeated measures techniques are used when you test the *same people* on more than one occasion, or you have matched pairs.
- Between-groups or independent-samples techniques are used when the subjects in each group are *different people* (or independent).
- One-way analysis of variance is used when you have only one *independent* variable.
- Two-way analysis of variance is used when you have two *independent* variables.
- Multivariate analysis of variance is used when you have more than one *dependent* variable.
- Analysis of covariance (ANCOVA) is used when you need to *control* for an additional variable which may be influencing the relationship between your independent and dependent variable.

Before we begin to explore some of the techniques available, there are a number of common issues that need to be considered. These topics will be relevant to many of the chapters included in this part of the book, so you may need to refer back to this section as you work through each chapter.

Assumptions

Each of the tests in this section have a number of assumptions underlying their use. There are some general assumptions that apply to all of the parametric techniques discussed here (e.g. t-tests, analysis of variance), and additional assumptions associated with specific techniques. The general assumptions are presented in this section and the more specific assumptions are presented in the following chapters, as appropriate. You will need to refer back to this section when using any of the techniques presented in Part Five. For information on the procedures used to check for violation of assumptions, see Tabachnick and Fidell (2001, Chapter 4). For further discussion of the consequences of violating the assumptions, see Stevens (1996, Chapter 6) and Glass, Peckham and Sanders (1972).

Level of measurement

Each of these approaches assumes that the dependent variable is measured at the interval or ratio level, that is, using a continuous scale rather than discrete categories. Wherever possible when designing your study, try to make use of continuous, rather than categorical, measures of your dependent variable. This gives you a wider range of possible techniques to use when analysing your data.

Random sampling

The techniques covered in Part Five assume that the scores are obtained using a random sample from the population. This is often not the case in real-life research.

Independence of observations

The observations that make up your data must be independent of one another. That is, each observation or measurement must not be influenced by any other observation or measurement. Violation of this assumption, according to Stevens (1996, p. 238), is very serious. There are a number of research situations that may violate this assumption of independence. Examples of some such studies are described below (these are drawn from Stevens, 1996, p. 239; and Gravetter & Wallnau, 2000, p. 262):

• Studying the performance of students working in pairs or small groups. The behaviour of each member of the group influences all other group members, thereby violating the assumption of independence.

• Studying the TV watching habits and preferences of children drawn from the same family. The behaviour of one child in the family (e.g. watching Program A) is likely to influence all children in that family, therefore the observations are not independent.

• Studying teaching methods within a classroom and examining the impact on students' behaviour and performance. In this situation all students could be influenced by the presence of a small number of trouble-makers, therefore individual behavioural or performance measurements are not independent.

Any situation where the observations or measurements are collected in a group setting, or subjects are involved in some form of interaction with one another, should be considered suspect. In designing your study you should try to ensure that all observations are independent. If you suspect some violation of this assumption, Stevens (1996, p. 241) recommends that you set a more stringent alpha value (e.g. p<.01).

Normal distribution

It is assumed that the populations from which the samples are taken are normally distributed. In a lot of research (particularly in the social sciences), scores on the dependent variable are not nicely normally distributed. Fortunately, most of the techniques are reasonably 'robust' or tolerant of violations of this assumption. With large enough sample sizes (e.g. 30+), the violation of this assumption should not cause any major problems (see discussion of this in Gravetter & Wallnau, 2000, p. 302; Stevens, 1996, p. 242). The distribution of scores for each of your groups can be checked using histograms obtained as part of the **Descriptive Statistics, Explore** option of SPSS (see Chapter 6). For a more detailed description of this process, see Tabachnick and Fidell (2001, pp. 99–104).

Homogeneity of variance

Techniques in this section make the assumption that samples are obtained from populations of equal variances. This means that the variability of scores for each of the groups is similar. To test this, SPSS performs the Levene test for equality of variances as part of the t-test and analysis of variances analyses. The results are presented in the output of each of these techniques. Be careful in interpreting the results of this test: you are hoping to find that the test is *not* significant (i.e. a significance level of *greater* than .05). If you obtain a significance value of less than .05, this suggests that variances for the two groups are not equal, and you have therefore violated the assumption of homogeneity of variance. Don't panic if you find this to be the case. Analysis of variance is reasonably robust to violations of this assumption, provided the size of your groups is reasonably similar (e.g. largest/smallest=1.5, Stevens, 1996, p. 249). For t-tests you are provided with two sets of results, for situations where the assumption is not violated and for when it is violated. In this case, you just consult whichever set of results is appropriate for your data.

Type 1 error, Type 2 error and power

As you will recall from your statistics texts, the purpose of t-tests and analysis of variance is to test hypotheses. With these types of analyses there is always the possibility of reaching the wrong conclusion. There are two different errors that we can make. We may reject the null hypothesis when it is, in fact, true (this is referred to as a Type 1 error). This occurs when we think there is a difference between our groups, but there really isn't. We can minimise this possibility by selecting an appropriate alpha level (the two levels often used are .05, and .01).

There is also a second type of error that we can make (Type 2 error). This occurs when we fail to reject a null hypothesis when it is, in fact, false (i.e. believing that the groups do not differ, when in fact they do). Unfortunately these two errors are inversely related. As we try to control for a Type 1 error, we actually increase the likelihood that we will commit a Type 2 error.

Ideally we would like the tests that we use to correctly identify whether in fact there is a difference between our groups. This is called the *power* of a test. Tests vary in terms of their power (e.g. parametric tests such as t-tests, analysis of variance etc. are more powerful than non-parametric tests); however, there are other factors that can influence the power of a test in a given situation:

- sample size;
- effect size (the strength of the difference between groups, or the influence of the independent variable); and
- alpha level set by the researcher (e.g. .05/.01).

The power of a test is very dependent on the size of the sample used in the study. According to Stevens (1996), when the sample size is large (e.g. 100 or more subjects), then 'power is not an issue' (p. 6). However, when you have a study where the group size is small (e.g. n=20), then you need to be aware of the possibility that a non-significant result may be due to insufficient power. Stevens (1996) suggests that when small group sizes are involved it may be necessary to adjust the alpha level to compensate (e.g. set a cut-off of .10 or .15, rather than the traditional .05 level).

There are tables available (see Cohen, 1988) that will tell you how large your sample size needs to be to achieve sufficient power, given the effect size you wish to detect. Ideally you would want an 80 per cent chance of detecting a relationship (if in fact one did exist). Some of the SPSS programs also provide an indication of the power of the test that was conducted, taking into account effect size and sample size. If you obtain a non-significant result and are using quite a small sample size, you need to check these power values. If the power of the test is less than .80 (80 per cent chance of detecting a difference), then you would need to interpret the reason for your nonsignificant result carefully. This may suggest insufficient power of the test, rather than no real difference between your groups. The power analysis gives an indication of how much confidence you should have in the results when you fail to reject the null hypothesis. The higher the power, the more confident you can be that there is no real difference between the groups.

Planned comparisons/Post-hoc analyses

When you conduct analysis of variance you are determining whether there are significant differences among the various groups or conditions. Sometimes you may be interested in knowing if, overall, the groups differ (that your independent variable in some way influences scores on your dependent variable). In other research contexts, however, you might be more focused and interested in testing the differences between specific groups, not between all the various groups. It is important that you distinguish which applies in your case, as different analyses are used for each of these purposes.

Planned comparisons (also know as *a priori*) are used when you wish to test specific hypotheses (usually drawn from theory or past research) concerning the differences between a subset of your groups (e.g. do Groups 1 and 3 differ significantly?). These comparisons need to be specified, or planned, before you analyse your data, not after fishing around in your results to see what looks interesting!

Some caution needs to be exercised with this approach if you intend to specify a lot of different comparisons. Planned comparisons do not control for the increased risks of Type 1 errors. A Type 1 error involves rejecting the null hypothesis (e.g. there are no differences among the groups) when it is actually true. In other words there is an increased risk of thinking that you have found a significant result when in fact it could have occurred by chance. If there are a large number of differences that you wish to explore, it may be safer to use the alternative approach (post-hoc comparisons), which is designed to protect against Type 1 errors.

The other alternative is to apply what is known as a Bonferroni adjustment to the alpha level that you will use to judge statistical significance. This involves setting a more stringent alpha level for each comparison, to keep the alpha across all the tests at a reasonable level. To achieve this you can divide your alpha level (usually .05) by the number of comparisons that you intend to make, and then use this new value as the required alpha level. For example, if you intend to make three comparisons the new alpha level would be .05 divided by 3 which equals .017. For a discussion on this technique, see Tabachnick and Fidell (2001, p. 50)

Post-hoc comparisons (also known as *a posteriori*) are used when you want to conduct a whole set of comparisons, exploring the differences between each of the groups or conditions in your study. If you choose this approach your analysis consists of two steps. First, an overall F ratio is calculated which tells you whether there are any significant differences among the groups in your design. If your overall F ratio is significant (indicating that there are differences among your groups), you can then go on and perform additional tests to identify where these differences occur (e.g. does Group 1 differ from Group 2 or Group 3, do Group 2 and Group 3 differ).

Post-hoc comparisons are designed to guard against the possibility of an increased Type 1 error due to the large number of different comparisons being made. This is done by setting more stringent criteria for significance, and therefore it is often harder to achieve significance. With small samples this can be a problem, as it can be very hard to find a significant result, even when the apparent difference in scores between the groups is quite large.

There are a number of different post-hoc tests that you can use, and these vary in terms of their nature and strictness. The assumptions underlying the post-hoc tests also differ. Some assume equal variances for the two groups (e.g. Tukey), others do not assume equal variance (e.g. Dunnett's C test). Two of the most commonly used post-hoc tests are Tukey's Honestly Significant Different test (HSD) and the Scheffe test. Of the two, the Scheffe test is the most cautious method for reducing the risk of a Type 1 error. However, the cost here is power. You may be less likely to detect a difference between your groups using this approach.

Effect size

All of the techniques discussed in this section will give you an indication of whether the difference between your groups is 'statistically significant' (i.e. not likely to have occurred by chance). It is typically a moment of great excitement for most researchers and students when they find their results are 'significant'! Unfortunately there is more to research than just obtaining statistical significance. What the probability values do not tell you is the degree to which the two variables are associated with one another. With large samples, even very small differences between groups can become statistically significant. This does not mean that the difference has any practical or theoretical significance.

One way that you can assess the importance of your finding is to calculate the 'effect size' (also known as 'strength of association'). This is a set of statistics which indicates the relative magnitude of the differences between means. In other words, it describes the 'amount of the total variance in the dependent variable that is predictable from knowledge of the levels of the independent variable' (Tabachnick & Fidell, 2001, p. 52).

There are a number of different effect size statistics, the most common of which are eta squared, Cohen's d and Cohen's f. SPSS calculates eta squared for you as part of the output from some techniques (analysis of variance). It does not provide eta squared for t-tests, but the formula and procedure that you can use to calculate it yourself is provided in Chapter 15.

Eta squared represents the proportion of variance of the dependent variable that is explained by the independent variable. Values for eta squared can range from 0 to 1. To interpret the strength of eta squared values the following guidelines can be used (from Cohen, 1988):

- .01=small effect;
- .06=moderate effect; and
- .14=large effect.

A number of criticisms have been levelled at eta squared (see Tabachnick & Fidell, 2001, pp. 52–53 for a discussion on this point). An alternative measure (partial eta squared) is available which overcomes a number of the concerns raised. This is the statistic calculated by SPSS (although it is not labelled as such in the output).

Missing data

When you are doing research, particularly with human beings, it is very rare that you will obtain complete data from every case. It is thus important that you inspect your data file for missing data. Run **Descriptives** and find out what percentage of values is missing for each of your variables. If you find a variable with a lot of unexpected missing data you need to ask yourself why. You should also consider whether your missing values are happening randomly, or whether there is some systematic pattern (e.g. lots of women failing to answer the question about their age). SPSS has a **Missing Value Analysis** procedure that may help find patterns in your missing values (see the bottom option under the **Analyze** menu).

You also need to consider how you will deal with missing values when you come to do your statistical analyses. The **Options** button in many of the SPSS statistical procedures offers you choices for how you want SPSS to deal with missing data. It is important that you choose carefully, as it can have dramatic effects on your results. This is particularly important if you are including a list of variables, and repeating the same analysis for all variables (e.g., correlations among a group of variables, t-tests for a series of dependent variables).

- The *Exclude cases listwise* option will include cases in the analysis only if it has full data on *all of the variables* listed in your variables box for that case. A case will be totally excluded from all the analyses if it is missing even one piece of information. This can severely, and unnecessarily, limit your sample size.
- The *Exclude cases pairwise* (sometimes shown as *Exclude cases analysis by analysis*) option, however, excludes the cases (persons) only if they are missing the data required for the specific analysis. They will still be included in any of the analyses for which they have the necessary information.
- The *Replace with mean* option, which is available in some SPSS statistical procedures, calculates the mean value for the variable and gives every missing case this value. This option should NEVER be used, as it can severely distort the results of your analysis, particularly if you have a lot of missing values.

Always press the **Options** button for any statistical procedure you conduct and check which of these options is ticked (the default option varies across procedures). I would strongly recommend that you use pairwise exclusion of missing data, unless you have a pressing reason to do otherwise. The only situation where you might need to use listwise exclusion is when you want to refer only to a subset of cases that provided a full set of results.

References

The chapters in Part Five assume that you have a good understanding of t-tests and analysis of variance. Some suggested readings are given below.

Cohen, J. (1988). *Statistical power analysis for the behavioral sciences*. Hillsdale, NJ: Erlbaum.

Cooper, D. R., & Schindler, P. S. (2003). *Business research methods* (8th edn). Boston: McGraw Hill.

Glass, G. V., Peckham, P. D., & Sanders, J. R. (1972). Consequences of failure to meet the assumptions underlying the use of analysis of variance and covariance. *Review of Educational Research, 42*, 237–288.

Gravetter, F. J., & Wallnau, L. B. (2000). *Statistics for the behavioral sciences* (5th edn). Belmont, CA: Wadsworth.

Hair, J. F., Anderson, R. E., Tatham, R. L., & Black, W. C. (1998). *Multivariate data analysis* (5th edn). Upper Saddle River, NJ: Prentice Hall.

Harris, R. J. (1994). *ANOVA: An analysis of variance primer*. Itasca, IL: Peacock.

Pagano, R. R. (1998). *Understanding statistics in the behavioral sciences* (5th edn). Pacific Grove, CA: Brooks/Cole.

Runyon, R. P., Coleman, K. A., & Pittenger, D. J. (2000). *Fundamentals of Behavioral Statistics* (9th edn). Boston: McGraw Hill.

Stevens, J. (1996). *Applied multivariate statistics for the social sciences* (3rd edn). Mahway, NJ: Lawrence Erlbaum.

Tabachnick, B. G., & Fidell, L. S. (2001). *Using multivariate statistics* (4th edn). New York: HarperCollins.

16 T-tests

There are a number of different types of t-tests available in SPSS. The two that will be discussed here are:

- *independent-samples t-test*, used when you want to compare the mean scores of two *different* groups of people or conditions; and
- *paired-samples t-test*, used when you want to compare the mean scores for the *same* group of people on two different occasions, or when you have matched pairs.

In both cases you are comparing the values on some continuous variable for *two* groups, or on *two* occasions. If you have more than two groups, or conditions, you will need to use analysis of variance instead.

For both of the t-tests discussed in this chapter there are a number of assumptions that you will need to check before conducting these analyses. The general assumptions common to both types of t-test are presented in the introduction to Part Five. The paired-samples t-test has an additional unique assumption that should also be checked. This is detailed in the Summary for Paired Samples T-tests. Before proceeding with the remainder of this chapter, you should read through the introduction to Part Five of this book.

Independent-samples t-test

An independent-samples t-test is used when you want to compare the mean score, on some *continuous* variable, for *two* different groups of subjects.

Details of example

To illustrate the use of this technique, the survey.sav data file (which is provided on the website accompanying this book; see p. xi) will be used. This example explores sex differences in self-esteem scores. The two variables used are SEX (with males coded as 1, and females coded as 2) and TSLFEST, which is the total score that participants recorded on a ten-item self-esteem scale (see the Appendix for more details on the study, the variables, and the questionnaire that was used to collect the data). If you would like to follow along with the steps detailed below you should start SPSS and open the survey.sav file now. This file can be opened only in SPSS.

Summary for independent-samples t-test

Example of research question:	Is there a significant difference in the mean self-esteem scores for males and females?
What you need:	Two variables:
	• one categorical, independent variable (e.g. males/females); and
	• one continuous, dependent variable (e.g. self-esteem scores).
What it does:	An independent-samples t-test will tell you whether there is a statistically significant difference in the mean scores for the two groups (that is, whether males and females differ significantly in terms of their self-esteem levels).
	In statistical terms, you are testing the probability that the two sets of scores (for males and females) came from the same population.
Assumptions:	The assumptions for this test are covered in the introduction to Part Five. You should read through that section before proceeding.
Non-parametric alternative:	Mann-Whitney Test (see Chapter 22).

Procedure for independent-samples t-test

1. From the menu at the top of the screen click on: **Analyze**, then click on **Compare means**, then on **Independent Samples T-test**.

2. Move the dependent (continuous) variable (e.g. total self-esteem) into the area labelled **Test variable**.

3. Move the independent variable (categorical) variable (e.g. sex) into the section labelled **Grouping variable**.

4. Click on **Define groups** and type in the numbers used in the data set to code each group. In the current data file 1=males, 2=females; therefore, in the **Group 1** box, type 1; and in the **Group 2** box, type 2.

5. Click on **Continue** and then **OK**.

The output generated from this procedure is shown below.

Group Statistics

	SEX	N	Mean	Std. Deviation	Std. Error Mean
Total self-esteem	MALES	184	34.02	4.91	.36
	FEMALES	252	33.17	5.71	.36

Independent Samples Test

		Levene's Test for Equality of Variances		t-test for Equality of Means						95% Confidence Interval of the Difference	
		F	Sig.	t	df	Sig. (2-tailed)	Mean Difference	Std. Error Difference	Lower	Upper
Total self-esteem	Equal variances assumed	3.506	.062	1.622	434	.105	.85	.52	-.18	1.87
	Equal variances not assumed			1.661	422.349	.098	.85	.51	-.16	1.85

Interpretation of output from independent-samples t-test

Step 1: Checking the information about the groups
In the **Group Statistics** box SPSS gives you the mean and standard deviation for each of your groups (in this case: male/female). It also gives you the number of people in each group (N). Always check these values first. Do they seem right? Are the N values for males and females correct? Or are there a lot of missing data? If so, find out why. Perhaps you have entered the wrong code for males and females (0 and 1, rather than 1 and 2). Check with your codebook.

Step 2: Checking assumptions
The first section of the **Independent Samples Test** output box gives you the results of Levene's test for equality of variances. This tests whether the variance (variation) of scores for the two groups (males and females) is the same. The outcome of this test determines which of the t-values that SPSS provides is the correct one for you to use.

- If your Sig. value is larger than .05 (e.g. .07, .10), you should use the first line in the table, which refers to **Equal variances assumed**.
- If the significance level of Levene's test is p=.05 or less (e.g. .01, .001), this means that the variances for the two groups (males/females) are *not* the same. Therefore your data violate the assumption of equal variance. Don't panic—SPSS is very kind and provides you with an alternative t-value which compensates for the fact that your variances are not the same. You should use the information in the *second* line of the t-test table, which refers to **Equal variances not assumed**.

Strange looking numbers

In your output you might come across strange looking numbers which take the form: 1.24E–02. These numbers have been written in scientific notation. They can easily be converted to a 'normal' number for presentation in your results. The number after the E tells you how many places you need to shift the decimal point. If the number after the E is negative, shift the decimal point to the left; if it is positive, shift it to the right. Therefore, 1.24E–02 becomes 0.124. The number 1.24E02 would be written as 124.

You can prevent these small numbers from being displayed in this strange way. Go to **Edit, Options** and put a tick in the box labelled: **No scientific notation for small numbers in tables**. Rerun the analysis and normal numbers should appear in all your output.

In the example given in the output above, the significance level for Levene's test is .06. This is larger than the cut-off of .05. This means that the assumption of equal variances has not been violated; therefore, when you report your t-value, you will use the one provided in the first line of the table.

Step 3: Assessing differences between the groups

To find out whether there is a significant difference between your two groups, refer to the column labelled **Sig. (2-tailed)**, which appears under the section labelled t-test for equality of means. Two values are given. One for equal variance, the other for unequal variance. Choose whichever your Levene's test result says you should use (see Step 2 above).

- If the value in the **Sig. (2-tailed)** column is *equal or less* than .05 (e.g. .03, .01, .001), then there is a significant difference in the mean scores on your dependent variable for each of the two groups.
- If the value is *above* .05 (e.g. .06, .10), there is no significant difference between the two groups.

In the example presented in the output above the Sig. (2-tailed) value is .105. As this value is *above* the required cut-off of .05, you conclude that there is *not* a statistically significant difference in the mean self-esteem scores for males and females.

Calculating the effect size for independent-samples t-test

In the introduction to Part Five of this book the issue of effect size was discussed. Effect size statistics provide an indication of the magnitude of the differences between your groups (not just whether the difference could have occurred by chance). There are a number of different effect size statistics, the most commonly used being eta squared. Eta squared can range from 0 to 1 and represents the proportion of variance in the dependent variable that is explained by the independent (group) variable. SPSS does not provide eta squared values for t-tests. It can, however, be calculated using the information provided in the output. The procedure for calculating eta squared is provided below.

The formula for eta squared is as follows:

$$\text{Eta squared} = \frac{t^2}{t^2 + (N1 + N2 - 2)}$$

Replacing with the appropriate values from the example above:

$$\text{Eta squared} = \frac{1.62^2}{1.62^2 + (184 + 252 - 2)}$$

$$\text{Eta squared} = .006$$

The guidelines (proposed by Cohen, 1988) for interpreting this value are: .01=small effect, .06=moderate effect, .14=large effect. For our current example you can see that the effect size of .006 is very small. Expressed as a percentage (multiply your eta square value by 100), only .6 per cent of the variance in self-esteem is explained by sex.

Presenting the results for independent-samples t-test

The results of the analysis could be presented as follows:

An independent-samples t-test was conducted to compare the self-esteem scores for males and females. There was no significant difference in scores for males (M=34.02, SD=4.91) and females [M=33.17, SD=5.71; t(434)=1.62, p=.11]. The magnitude of the differences in the means was very small (eta squared=.006).

Paired-samples t-test

Paired-samples t-test (also referred to as repeated measures) is used when you have only one group of people (or companies, or machines etc.) and you collect data from them on two different occasions, or under two different conditions. Pre-test/post-test experimental designs are an example of the type of situation where this technique is appropriate. You assess each person on some continuous measure at Time 1, and then again at Time 2, after exposing them to some experimental manipulation or intervention. This approach is used also when you have matched pairs of subjects (i.e. each person is matched with another on specific criteria, such as age, sex). One of the pair is exposed to Intervention 1 and the other is exposed to Intervention 2. Scores on a continuous measure are then compared for each pair.

Paired-samples t-tests can also be used when you measure the same person in terms of his/her response to two different questions (e.g. asking him/her to rate the importance in terms of life satisfaction on two dimensions of life: health, financial security). In this case, both dimensions should be rated on the same scale (e.g. from 1=not at all important to 10=very important).

Details of example

To illustrate the use of paired-samples t-test I will be using the data from the file labelled experim.sav (this is included on the website accompanying this book; see p. xi). This is a 'manufactured' data file—created and manipulated to illustrate a number of different statistical techniques. Full details of the study design, the measures used etc. are provided in the Appendix. For the example below I will be exploring the impact of an intervention designed to increase students' confidence in their ability to survive a compulsory statistics course. Students were asked to complete a Fear of Statistics Test (FOST) both before (Time 1) and after the intervention (Time 2). The two variables from the data file that I will be using are: FOST1 (scores on the Fear of Statistics Test at Time 1) and FOST2 (scores on the Fear of Statistics Test at Time 2). If you wish to follow along with the following steps you should start SPSS and open the file labelled experim.sav. This file can be opened only in SPSS.

Summary for paired-samples t-test

Example of research question: Is there a significant change in participants' fear of statistics scores following participation in an intervention designed to increase students' confidence in their ability to successfully complete a statistics course?

Does the intervention have an impact on participants' fear of statistics scores?

What you need: One set of subjects (or matched pairs). Each person (or pair) must provide both sets of scores.

Two variables: • one categorical independent variable (in this case it is Time: with two different levels Time 1, Time 2); and
• one continuous, dependent variable (e.g. Fear of Statistics Test scores) measured on two different occasions, or under different conditions.

What it does: A paired-samples t-test will tell you whether there is a statistically significant difference in the mean scores for Time 1 and Time 2.

Assumptions: The basic assumptions for t-tests are covered in the introduction to Part Five. You should read through that section before proceeding.
Additional assumption: The difference between the two scores obtained for each subject should be normally distributed. With sample sizes of 30+, violation of this assumption is unlikely to cause any serious problems.

Non-parametric alternative: Wilcoxon Signed Rank Test (see Chapter 22).

Procedure for paired-samples t-test

1. From the menu at the top of the screen click on: **Analyze**, then click on **Compare Means**, then on **Paired Samples T-test**.

2. Click on the two variables that you are interested in comparing for each subject (e.g. fost1: fear of stats time1, fost2: fear of stats time2).

3. With both of the variables highlighted, move them into the box labelled **Paired Variables** by clicking on the arrow button. Click on **OK**.

The output generated from this procedure is shown below

Paired Samples Statistics

		Mean	N	Std. Deviation	Std. Error Mean
Pair 1	fear of stats time1	40.17	30	5.16	.94
	fear of stats time2	37.50	30	5.15	.94

Paired Samples Test

		Paired Differences							
					95% Confidence Interval of the Difference				
		Mean	Std. Deviation	Std. Error Mean	Lower	Upper	t	df	Sig. (2-tailed)
Pair 1	fear of stats time1 - fear of stats time2	2.67	2.71	.49	1.66	3.68	5.394	29	.000

Interpretation of output from paired-samples t-test

There are two steps involved in interpreting the results of this analysis.

Step 1: Determining overall significance

In the table labelled **Paired Samples Test** you need to look in the final column, labelled **Sig. (2-tailed)**—this is your probability value. If this value is less than .05 (e.g. .04, .01, .001), then you can conclude that there is a significant difference between your two scores. In the example given above the probability value is .000. This has actually been rounded down to three decimal places—it means that the actual probability value was less than .0005. This value is substantially smaller than our specified alpha value of .05. Therefore, we can conclude that there is a significant difference in the Fear of Statistics Test scores at Time 1 and at Time 2. Take note of the t-value (in this case, 5.39) and the degrees of freedom (df=29), as you will need these when you report your results.

Step 2: Comparing mean values

Having established that there is a significant difference, the next step is to find out which set of scores is higher (Time 1 or Time 2). To do this, look in the first printout box, labelled **Paired Samples Statistics**. This box gives you the Mean scores for each of the two sets of scores. In our case, the mean Fear of Stats score at Time 1 was 40.17 and the mean score at Time 2 was 37.50. Therefore, we can conclude that there was a significant decrease in Fear of Statistics Test scores from Time 1 (prior to the intervention) to Time 2 (after the intervention).

Calculating the effect size for paired-samples t-test

Although the results presented above tell us that the difference we obtained in the two sets of scores was unlikely to occur by chance, it does not tell us much about the magnitude of the intervention's effect. One way to do this is to calculate an effect size statistic (see the introduction to Part Five for more on this point). The procedure for calculating and interpreting eta squared (one of the most commonly used effect size statistics) is presented below.

Eta squared can be obtained using the following formula:

$$\text{Eta squared} = \frac{t^2}{t^2 + N - 1}$$

Replace with the values from the example presented above:

$$\text{Eta squared} = \frac{(5.39)^2}{(5.39)^2 + 30 - 1}$$
$$= \frac{29.05}{29.05 + 30 - 1}$$
$$= .50$$

To interpret the eta squared values the following guidelines can be used (from Cohen, 1988): .01=small effect, .06=moderate effect, .14=large effect. Given our eta squared value of .50, we can conclude that there was a large effect, with a substantial difference in the Fear of Statistics scores obtained before and after the intervention.

Caution

Although we obtained a significant difference in the scores before/after the intervention, we cannot say that the intervention *caused* the drop in Fear of Statistics Test scores. Research is never that simple, unfortunately! There are many other factors that may have also influenced the decrease in fear scores. Just the passage of time (without any intervention) could have contributed. Any number of other events may also have occurred during this period that influenced students' attitudes to statistics. Perhaps the participants were exposed to previous statistics students who told them how great the instructor was and how easy it was to pass the course. Perhaps they were all given an illegal copy of the statistics exam (with all answers included)! There are many other possible confounding or contaminating factors.

Wherever possible, the researcher should try to anticipate these confounding factors and either control for them or incorporate them into the research design. In the present case, the use of a control group that was not exposed to an intervention but was similar to the participants in all other ways would have improved the study. This would have helped to rule out the effects of time, other events etc. that may have influenced the results of the current study.

Presenting the results for paired-samples t-test

The key details that need to be presented are the name of the test, the purpose of the test, the t-value, the degrees of freedom (df), the probability value, and the means and standard deviations for each of the groups or administrations. It is also a good idea to present an effect size statistic (e.g. eta squared).
The results of the analysis conducted above could be presented as follows:

A paired-samples t-test was conducted to evaluate the impact of the intervention on students' scores on the Fear of Statistics Test (FOST). There was a statistically significant decrease in FOST scores from Time 1 (*M*=40.17, *SD*=5.16) to Time 2 [*M*=37.5, *SD*=5.15, *t*(29)=5.39, *p*<.0005]. The eta squared statistic (.50) indicated a large effect size.

Additional exercises

Business

Data file: *staffsurvey.sav*. See Appendix for details of the data file.

1. Follow the procedures in the section on independent samples t-tests to compare the mean staff satisfaction scores (*totsatis*) for permanent and casual staff (*employstatus*). Is there a significant difference in mean satisfaction scores?

Health

Data file: *sleep.sav*. See Appendix for details of the data file.

1. Follow the procedures in the section on independent samples t-tests to compare the mean sleepiness ratings (Sleepiness and Associated Sensations Scale total score: *totSAS*) for males and females (*gender*). Is there a significant difference in mean sleepiness scores?

Reference

Cohen, J. (1988). *Statistical power analysis for the behavioral sciences*. Hillsdale, NJ: Erlbaum.

17 One-way analysis of variance

In the previous chapter we used t-tests to compare the scores of two different groups or conditions. In many research situations, however, we are interested in comparing the mean scores of more than two groups. In this situation we would use analysis of variance (ANOVA). One-way analysis of variance involves one independent variable (referred to as a factor), which has a number of different levels. These levels correspond to the different groups or conditions. For example, in comparing the effectiveness of three different teaching styles on students' Maths scores, you would have one factor (teaching style) with three levels (e.g. whole class, small group activities, self-paced computer activities). The dependent variable is a continuous variable (in this case, scores on a Maths test).

Analysis of variance is so called because it compares the variance (variability in scores) *between* the different groups (believed to be due to the independent variable) with the variability *within* each of the groups (believed to be due to chance). An F ratio is calculated which represents the variance between the groups, divided by the variance within the groups. A large F ratio indicates that there is more variability between the groups (caused by the independent variable) than there is within each group (referred to as the error term).

A significant F test indicates that we can reject the null hypothesis, which states that the population means are equal. It does not, however, tell us which of the groups differ. For this we need to conduct post-hoc tests. The alternative to conducting post-hoc tests after obtaining a significant 'omnibus' F test is to plan your study to conduct only specific comparisons (referred to as planned comparisons).

A comparison of post-hoc versus planned comparisons is presented in the introduction to Part Five of this book (other suggested readings are also provided). There are advantages and disadvantages to each approach—you should consider your choice carefully, before beginning your analysis. Post-hoc tests are designed to help protect against the likelihood of a Type 1 error, but this approach is stricter, making it more difficult to obtain statistically significant differences. Unless you have clear conceptual grounds for wishing only to compare specific groups, then it may be more appropriate to use post-hoc analysis.

In this chapter two different types of one-way ANOVA are discussed:

- between-groups analysis of variance, which is used when you have different subjects or cases in each of your groups (this is referred to as an independent-groups design); and

- repeated-measures analysis of variance, which is used when you are measuring the same subjects under different conditions (or measured at different points in time) (this is also referred to as a within-subjects design).

In the between-groups analysis of variance section that follows, the use of both post-hoc tests and planned comparisons will be illustrated.

One-way between-groups ANOVA with post-hoc tests

One-way between-groups analysis of variance is used when you have one independent (grouping) variable with three or more levels (groups) and one dependent continuous variable. The 'one-way' part of the title indicates there is only one independent variable, and 'between-groups' means that you have different subjects or cases in each of the groups.

Details of example

To demonstrate the use of this technique I will be using the survey.sav data file included on the website accompanying this book (see p. xi). The data come from a survey that was conducted to explore the factors that affect respondents' psychological adjustment, health and wellbeing. This is a real data file from actual research conducted by a group of my Graduate Diploma students. Full details of the study, the questionnaire and scales used are provided in the Appendix. If you wish to follow along with the steps described in this chapter, you should start SPSS and open the file labelled survey.sav. This file can be opened only in SPSS.

Details of the variables used in this analysis are provided in the table below:

File name	Variable name	Variable label	Coding instructions
survey.sav	Toptim	Total optimism	Total score on the Optimism scale. Scores can range from 6 to 30 with high scores indicating higher levels of optimism.
	Agegp3	Agegp3	This variable is a recoded variable, dividing age into three equal groups (see instructions for how to do this in Chapter 8): Group 1: 18–29 = 1 Group 2: 30–44 = 2 Group 3: 45+ = 3

Summary for one-way between-groups ANOVA with post-hoc tests

Example of research question: Is there a difference in optimism scores for young, middle-aged and old subjects?

What you need:	Two variables: • one categorical independent variable with three or more distinct categories. This can also be a continuous variable that has been recoded to give three equal groups (e.g. age group: subjects divided into 3 age categories, 29 and younger, between 30 and 44, 45 or above). For instructions on how to do this see Chapter 8; and • one continuous dependent variable (e.g. optimism).
What it does:	One-way ANOVA will tell you whether there are significant differences in the mean scores on the dependent variable across the three groups. Post-hoc tests can then be used to find out where these differences lie.
Assumptions:	See discussion of the general ANOVA assumptions presented in the introduction to Part Five.
Non-parametric alternative:	Kruskal-Wallis Test (see Chapter 22).

Procedure for one-way between-groups ANOVA with post-hoc tests

1. From the menu at the top of the screen click on: **Analyze**, then click on **Compare Means**, then on **One-way ANOVA**.

2. Click on your dependent (continuous) variable (e.g. Total optimism). Move this into the box marked **Dependent List** by clicking on the arrow button.

3. Click on your independent, categorical variable (e.g. agegp3). Move this into the box labelled **Factor**.

4. Click the **Options** button and click on **Descriptive**, **Homogeneity of variance test**, **Brown-Forsythe**, **Welsh** and **Means Plot**.

5. For **Missing values**, make sure there is a dot in the option marked **Exclude cases analysis by analysis**. If not, click on this option once. Click on Continue.

6. Click on the button marked **Post Hoc**. Click on **Tukey**.

7. Click on **Continue** and then **OK**.

The output generated from this procedure is shown below.

Descriptives

Total Optimism

	N	Mean	Std. Deviation	Std. Error	95% Confidence Interval for Mean Lower Bound	95% Confidence Interval for Mean Upper Bound	Minimum	Maximum
18-29	147	21.36	4.55	.38	20.62	22.10	7	30
30-44	153	22.10	4.15	.34	21.44	22.77	10	30
45+	135	22.96	4.49	.39	22.19	23.72	8	30
Total	435	22.12	4.43	.21	21.70	22.53	7	30

Test of Homogeneity of Variances

Total Optimism

Levene Statistic	df1	df2	Sig.
.746	2	432	.475

ANOVA

Total Optimism

	Sum of Squares	df	Mean Square	F	Sig.
Between Groups	179.069	2	89.535	4.641	.010
Within Groups	8333.951	432	19.292		
Total	8513.021	434			

Robust Tests of Equality of Means

total optimism

	Statistic[a]	df1	df2	Sig.
Welch	4.380	2	284.508	.013
Brown-Forsythe	4.623	2	423.601	.010

a. Asymptotically F distributed.

Multiple Comparisons

Dependent Variable: Total Optimism

Tukey HSD

(I) AGEGP3	(J) AGEGP3	Mean Difference (I-J)	Std. Error	Sig.	95% Confidence Interval Lower Bound	Upper Bound
18-29	30-44	-.74	.51	.307	-1.93	.44
	45+	-1.60*	.52	.007	-2.82	-.37
30-44	18-29	.74	.51	.307	-.44	1.93
	45+	-.85	.52	.229	-2.07	.36
45+	18-29	1.60*	.52	.007	.37	2.82
	30-44	.85	.52	.229	-.36	2.07

* The mean difference is significant at the .05 level.

Means plot

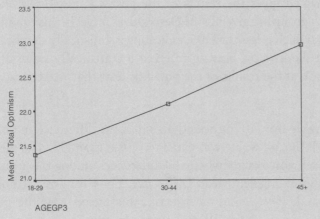

Interpretation of output from one-way between-groups ANOVA with post-hoc tests

One-way ANOVA, with the additional options checked above, will give you five separate bits of information, as follows:

Descriptives

This table gives you information about each group (number in each group, means, standard deviation, minimum and maximum, etc.) Always check this table first. Are the Ns for each group correct?

Test of homogeneity of variances

The homogeneity of variance option gives you Levene's test for homogeneity of variances, which tests whether the variance in scores is the same for each of the three groups. Check the significance value (Sig.) for Levene's test. If this number is *greater* than .05 (e.g. .08, .12, .28), then you have *not* violated the assumption of homogeneity of variance. In this example the Sig. value is .475. As this is greater than .05, we have not violated the homogeneity of variance assumption. If you have found that you violated this assumption you will need to consult the table in the output headed **Robust Tests of Equality of Means**. The two tests shown there (Welsh and Brown-Forsythe) are preferable when the assumption of the homogeneity of variance is violated.

ANOVA

This table gives both between-groups and within-groups sums of squares, degrees of freedom etc. You will recognise these from your statistics books. The main thing you are interested in is the column marked **Sig**. If the Sig. value is less than or equal to .05 (e.g. .03, .01, .001), then there is a significant difference somewhere among the mean scores on your dependent variable for the three groups. This does not tell you which group is different from which other group. The statistical significance of the differences between each pair of groups is provided in the table labelled **Multiple Comparisons**, which gives the results of the post-hoc tests (described below). The means for each group are given in the **Descriptives** table. In this example the overall Sig. value is .01, which is less than .05, indicating a statistically significant result somewhere among the groups. Having received a statistically significant difference, we can now look at the results of the post-hoc tests that we requested.

Multiple comparisons

You should look at this table only if you found a significant difference in your overall ANOVA. That is, if the Sig. value was equal to or less than .05. The post-hoc tests in this table will tell you exactly where the differences among the groups occur. Look down the column labelled **Mean Difference**. Look for any asterisks (*) next to the values listed. If you find an asterisk, this means that the two groups being compared are significantly different from one another at the p<.05

level. The exact significance value is given in the column labelled **Sig**. In the results presented above, only group 1 and group 3 are statistically significantly different from one another. That is, the 18–29 age group and the 45+ age group differ significantly in terms of their optimism scores.

Means plots

This plot provides an easy way to compare the mean scores for the different groups. You can see from this plot that the 18–29 age group recorded the lowest optimism scores with the 45+ age group recording the highest.

Warning: these plots can be misleading. Depending on the scale used on the Y axis (in this case representing Optimism scores), even small differences can look dramatic. In the above example the actual difference in the mean scores between the groups is very small (21.36, 22.10, 22.96), while on the graph it looks substantial. The lesson here is: don't get too excited about your plots until you have compared the mean values (available in the Descriptives box) and the scale used in the plot.

Calculating effect size

Although SPSS does not generate it for this analysis, it is possible to determine the effect size for this result (see the introduction to Part Five for a discussion on effect sizes). The information you need to calculate eta squared, one of the most common effect size statistics, is provided in the ANOVA table (a calculator would be useful here). The formula is:

$$\text{Eta squared} = \frac{\text{Sum of squares between-groups}}{\text{Total sum of squares}}$$

In this example all you need to do is to divide the Sum of squares for between-groups (179.07) by the Total sum of squares (8513.02). The resulting eta squared value is .02, which in Cohen's (1988) terms would be considered a small effect size. Cohen classifies .01 as a small effect, .06 as a medium effect and .14 as a large effect.

Warning

In this example we obtained a statistically significant result, but the actual difference in the mean scores of the groups was very small (21.36, 22.10, 22.96). This is evident in the small effect size obtained (eta squared=.02). With a large enough sample (in this case N=435), quite small differences can become statistically significant, even if the difference between the groups is of little practical importance. Always interpret your results carefully, taking into account all the information you have available. Don't rely too heavily on statistical significance—many other factors also need to be considered.

Presenting the results from one-way between-groups ANOVA with post-hoc tests

The results of the one-way between-groups analysis of variance with post-hoc tests could be presented as follows:

A one-way between-groups analysis of variance was conducted to explore the impact of age on levels of optimism, as measured by the Life Orientation test (LOT). Subjects were divided into three groups according to their age (Group 1: 29 or less; Group 2: 30 to 44; Group 3: 45 and above). There was a statistically significant difference at the $p<.05$ level in LOT scores for the three age groups [$F(2, 432)=4.6$, $p=.01$]. Despite reaching statistical significance, the actual difference in mean scores between the groups was quite small. The effect size, calculated using eta squared, was .02. Post-hoc comparisons using the Tukey HSD test indicated that the mean score for Group 1 ($M=21.36$, $SD=4.55$) was significantly different from Group 3 ($M=22.96$, $SD=4.49$). Group 2 ($M=22.10$, $SD=4.15$) did not differ significantly from either Group 1 or 3.

One-way between-groups ANOVA with planned comparisons

In the example provided above we were interested in comparing optimism scores across each of the three groups. In some situations, however, researchers may be interested only in comparisons between specific groups. For example, in an experimental study with five different interventions we may want to know whether Intervention 1 is superior to each of the other interventions. In that situation we may not be interested in comparing all the possible combinations of groups. If we are interested in only a subset of the possible comparisons it makes sense to use planned comparisons, rather than post-hoc tests, because of 'power' issues (see discussion of power in the introduction to Part Five). Planned comparisons are more 'sensitive' in detecting differences. Post-hoc tests, on the other hand, set more stringent significance levels to reduce the risk of a Type 1 error, given the larger number of tests performed. The choice of whether to use planned comparisons versus post-hoc tests must be made before you begin your analysis. It is not appropriate to try both and see which results you prefer!

To illustrate planned comparisons I will use the same data as the previous example. Normally you would not conduct both analyses. In this case we will consider a slightly different question: Are subjects in the older age group (45+yrs) more optimistic than those in the two younger age groups (18–29yrs; 30–44yrs)?

Specifying coefficient values

In the following procedure you will be asked by SPSS to indicate which particular groups you wish to compare. To do this you need to specify 'coefficient' values. Many students find this confusing initially, so I will explain this process here.

First, you need to identify your groups based on the different values of the independent variable (agegp3):

Group 1 (coded as 1)	Group 2 (coded as 2)	Group 3 (coded as 3)
Age 18–29	Age 30–44	Age 45+

Next, you need to decide which of the groups you wish to compare and which you wish to ignore. I will illustrate this process using a few examples.

Example 1

To compare Group 3 with the other two groups, the coefficients would be as follows:

Group 1	Group 2	Group 3
–1	–1	2

The values of the coefficients across the row should add up to 0. Coefficients with different values are compared. If you wished to ignore one of the groups you would give it a coefficient of 0. You would then need to adjust the other coefficient values so that they added up to 0.

Example 2

To compare Group 3 with Group 1 (ignoring Group 2), the coefficients would be as follows:

Group1	Group 2	Group 3
–1	0	1

This information on the coefficients for each of your groups will be required in the Contrasts section of the procedure that follows.

Procedure for one-way between-groups ANOVA with planned comparisons

1. From the menu at the top of the screen click on: **Analyze**, then click on **Compare Means**, then on **One-way ANOVA**.

2. Click on your dependent (continuous) variable (e.g. total optimism). Click on the arrow to move this variable into the box marked **Dependent List**.

3. Click on your independent, categorical variable (e.g. agegp3). Move this into the box labelled **Factor**.

4. Click the **Options** button and click on **Descriptive**, **Homogeneity-of-Variance** and **Means Plot**.

5. For **Missing Values**, make sure there is a dot in the option marked **Exclude cases analysis by analysis**. Click on **Continue**.

6. Click on the **Contrasts** button.
   - In the **Coefficients** box type the coefficient for the first group (from Example 1 above, this value would be –1). Click on **Add**.
   - Type in the coefficient for the second group (–1). Click on **Add**.
   - Type in the coefficient for the third group (2). Click on **Add**.
   - The **Coefficient Total** should be 0 if you have entered all the coefficients correctly.

7. Click on **Continue** and then **OK**.

The output generated from this procedure is shown below. Only selected output is displayed.

Contrast Coefficients

	AGEGP3		
Contrast	18-29	30-44	45+
1	-1	-1	2

Contrast Tests

		Contrast	Value of Contrast	Std. Error	t	df	Sig. (2-tailed)
Total Optimism	Assume equal variances	1	2.45	.91	2.687	432	.007
	Does not assume equal	1	2.45	.92	2.654	251.323	.008

Interpretation of output from one-way between-groups ANOVA with planned comparisons

The Descriptives and Test of homogeneity of variances tables generated as part of this output are the same as obtained in the previous example for one-way ANOVA with post-hoc tests. Only the output relevant to planned comparisons will be discussed here.

Step 1

In the table labelled **Contrast Coefficients,** the coefficients that you specified for each of your groups is provided. Check that this is what you intended.

Step 2

The main results that you are interested in are presented in the table labelled **Contrast Tests.** As we can assume equal variances (our Levene's test was not significant), we will use the first row in this table. The Sig. level for the contrast that we specified is .007. This is less than .05, so we can conclude that there is a statistically significant difference between Group 3 (45+ age group) and the other two groups. Although statistically significant, the actual difference between the mean scores of these groups is very small (21.36, 22.10, 22.96). Refer to the results of the previous section for further discussion on this point.

You will notice that the result of the planned comparisons analysis is expressed using a t statistic, rather than the usual F ratio associated with analysis of variance. To obtain the corresponding F value, all you need to do is to square the t value. In this example the t value is 2.687, which, when squared, equals 7.22. To report the results you also need the degrees of freedom. The first value (for all planned comparisons) is 1; the second is given in the table next to the t value (in this example the value is 432). Therefore, the results would be expressed as F (1, 432)=7.22, p=.007.

One-way repeated measures ANOVA

In a one-way repeated measures ANOVA design each subject is exposed to two or more different conditions, or measured on the same continuous scale on three or more occasions. It can also be used to compare respondents' responses to two or more different questions or items. These questions, however, must be measured using the same scale (e.g. 1=strongly disagree, to 5=strongly agree).

   The use of post-hoc tests and planned comparisons with a repeated measures design is beyond the scope of this book and will not be demonstrated here. For further reading on this topic, see Chapter 13 in Stevens (1996).

Details of example

To demonstrate the use of this technique, the data file labelled experim.sav (included on the website provided with this book; see p. xi) will be used. Details of this data file can be found in the Appendix. A group of students were invited to participate in an intervention designed to increase their confidence in their ability to do statistics. Their confidence levels (as measured by a self-report scale) were measured before the intervention (Time 1), after the intervention (Time 2) and again three months later (Time 3). If you wish to follow along with the procedure, you should start SPSS and open the file labelled experim.sav now. This file can be opened only in SPSS. Details of the variables names and labels from the data file are as follows:

File name	Variable name	Variable label	Coding instructions
experim.sav	Confid1	Confidence scores at time1	Total scores on the Confidence in Coping with Statistics test administered prior to the program. Scores range from 10 to 40. High scores indicate higher levels of confidence.
	Confid2	Confidence scores at time2	Total scores on the Confidence in Coping with Statistics test administered after the program. Scores range from 10 to 40. High scores indicate higher levels of confidence.
	Confid3	Confidence scores at time3	Total scores on the Confidence in Coping with Statistics test administered 3 months later. Scores range from 10 to 40. High scores indicate higher levels of confidence.

Summary for one-way repeated measures ANOVA

Example of research question: Is there a change in confidence scores over the three time periods?

What you need: One group of subjects measured on the same scale on three different occasions or under three different conditions. Or each person measured on three different questions or items (using the same response scale).

This involves two variables:
- one independent variable (categorical) (e.g. Time 1/ Time 2/ Time 3); and
- one dependent variable (continuous) (e.g. scores on the Confidence test).

What it does: This technique will tell you if there is a significant difference somewhere among the three sets of scores.

Assumptions: See discussion of the general ANOVA assumptions presented in the introduction to Part Five.

Non-parametric alternative: Friedman Test (see Chapter 22).

Procedure for one-way repeated measures ANOVA

1. From the menu at the top of the screen click on: **Analyze**, then click on **General Linear Model**, then on **Repeated Measures**.

2. In the **Within Subject Factor Name** box type in a name that represents your independent variable (e.g. Time or Condition). This is not an actual variable name, just a label you give your independent variable.

3. In the **Number of Levels** box type the number of levels or groups (time periods) involved (in this example it is 3).

4. Click **Add**.

5. Click on the **Define** button on the right-hand side.

6. Select the three variables that represent your repeated measures variable (e.g. confid1, confid2, confid3). Click on the arrow button to move them into the **Within Subjects Variables** box.

7. Click on the **Options** box at the bottom right of your screen.

8. Tick the **Descriptive Statistics** and **Estimates of effect size** boxes in the area labelled **Display**.

9. Click on **Continue** and then **OK**.

The output generated from this procedure is shown below.

Descriptive Statistics

	Mean	Std. Deviation	N
confidence time1	19.00	5.37	30
confidence time2	21.87	5.59	30
confidence time3	25.03	5.20	30

Multivariate Tests[b]

Effect		Value	F	Hypothesis df	Error df	Sig.	Partial Eta Squared
time	Pillai's Trace	.749	41.711[a]	2.000	28.000	.000	.749
	Wilks' Lambda	.251	41.711[a]	2.000	28.000	.000	.749
	Hotelling's Trace	2.979	41.711[a]	2.000	28.000	.000	.749
	Roy's Largest Root	2.979	41.711[a]	2.000	28.000	.000	.749

a. Exact statistic

b.
   Design: Intercept
   Within Subjects Design: time

Mauchly's Test of Sphericity[b]

Measure: MEASURE_1

Within Subjects Effect	Mauchly's W	Approx. Chi-Square	df	Sig.	Epsilon[a]		
					Greenhouse-Geisser	Huynh-Feldt	Lower-bound
TIME	.592	14.660	2	.001	.710	.737	.500

Tests the null hypothesis that the error covariance matrix of the orthonormalized transformed dependent variables is proportional to an identity matrix.

a. May be used to adjust the degrees of freedom for the averaged tests of significance. Corrected tests are displayed in the Tests of Within-Subjects Effects table.

b. Design: Intercept
   Within Subjects Design: TIME

Interpretation of output from one-way repeated measures ANOVA

You will notice that this technique generates a lot of complex-looking output. This includes tests for the assumption of sphericity, and both univariate and multivariate ANOVA results. Full discussion of the difference between the univariate and multivariate results is beyond the scope of this book; in this chapter only the multivariate results will be discussed (see Stevens, 1996, pp. 466–469 for more information).

The reason for interpreting the multivariate statistics provided by SPSS is that the univariate statistics make the assumption of sphericity. The sphericity assumption requires that the variance of the population difference scores for any two conditions are the same as the variance of the population difference scores for any other two conditions (an assumption that is commonly violated). This is assessed by SPSS using Mauchly's Test of Sphericity.

The multivariate statistics, however, do not require sphericity. You will see in our example that we have violated the assumption of sphericity, as indicated by the Sig. value of .001 in the box labelled **Mauchly's Test of Sphericity**. Although there are ways to compensate for this assumption violation, it is safer to inspect the multivariate statistics provided in the output.

Let's look at the key values in the output that you will need to consider.

Descriptive statistics

In the first output box you are provided with the descriptive statistics for your three sets of scores (Mean, Standard deviation, N). It is a good idea to check that these make sense. Are there the right number of people in each group? Do the mean values make sense given the scale that was used? In the example above you will see that the lowest mean confidence score was for Time 1 (before the intervention) and the highest at Time 3 (after the statistics course was completed).

Multivariate tests

In this table the value that you are interested in is **Wilks' Lambda,** also the associated probability value given in the column labelled **Sig.** All of the multivariate tests yield the same result, but the most commonly reported statistic is Wilks' Lambda. In this example the value for Wilks' Lambda is .25, with a probability value of .000 (which really means p<.0005). The p value is less than .05; therefore we can conclude that there is a statistically significant effect for time. This suggests that there was a change in confidence scores across the three different time periods.

Effect size

Although we have found a statistically significant difference between the three sets of scores, we also need to assess the effect size of this result (see discussion on effect sizes in the introduction to Part Five of this book). The value you are interested in is Partial Eta squared, given in the **Multivariate Tests** output box. The value obtained in this study is .749. Using the commonly used guidelines proposed by Cohen (1988) (.01=small, .06=moderate, .14=large effect), this result suggests a very large effect size.

If you obtain a statistically significant result from the above analyses this suggests that there is a difference somewhere among your groups. It does not tell you which groups or set of scores (in this case, Time 1, Time 2, Time 3) differ from one another. It is necessary to conduct further post-hoc tests to identify which groups differed. Discussion of these tests are beyond the scope of this book; however, I suggest that you consult Stevens (1996, Chapter 13).

Presenting the results from one-way repeated measures ANOVA

The results of a one-way repeated measures ANOVA could be presented as follows:

A one-way repeated measures ANOVA was conducted to compare scores on the Confidence in Coping with Statistics test at Time 1 (prior to the intervention), Time 2 (following the intervention) and Time 3 (three-month follow-up). The means and standard deviations are presented in Table XX. There was a significant effect for time [Wilks' Lambda=.25, $F(2, 28)=41.17$, $p<.0005$, multivariate partial eta squared=.75.]

Table XX
Descriptive Statistics for Confidence in Coping with Statistics Test Scores for Time 1, Time 2 and Time 3.

Time period	N	Mean	Standard deviation
Time 1 (Pre-intervention)	30	19.00	5.37
Time 2 (Post-intervention)	30	21.87	5.59
Time 3 (3-month follow-up)	30	25.03	5.20

Additional exercises

Business

Data file: *staffsurvey.sav*. See Appendix for details of the data file.

1. Conduct a one-way ANOVA with post hoc tests (if appropriate) to compare staff satisfaction scores (*totsatis*) across each of the length of service categories (use the *servicegp3* variable).

Health

Data file: *sleep.sav*. See Appendix for details of the data file.

1. Conduct a one-way ANOVA with post hoc tests (if appropriate) to compare the mean sleepiness ratings (Sleepiness and Associated Sensations Scale total score: *totSAS*) for the three age groups defined by the variable *agegp3* (<=37, 38–50, 51+).

References

Cohen, J. (1988). *Statistical power analysis for the behavioral sciences*. Hillsdale, NJ: Erlbaum.

Stevens, J. (1996). *Applied multivariate statistics for the social sciences* (3rd edn). Mahway, NJ: Lawrence Erlbaum.

18 Two-way between-groups ANOVA

In this section we will explore two-way, between-groups analysis of variance. *Two-way* means that there are two independent variables, and *between-groups* indicates that different people are in each of the groups. This technique allows us to look at the individual and joint effect of two independent variables on one dependent variable. In Chapter 17 we used one-way between-groups ANOVA to compare the optimism scores for three age groups (18–29, 30–44, 45+). We found a significant difference between the groups, with post-hoc tests indicating that the major difference was between the youngest and oldest groups. Older people reported higher levels of optimism.

The next question we can ask is: Is this the case for both males and females? One-way ANOVA cannot answer this question—the analysis was conducted on the sample as a whole, with males and females combined. In this chapter I will take the investigation a step further and consider the impact of gender on this finding. I will therefore have two independent variables (age group and sex), and one dependent variable (optimism).

The advantage of using a two-way design is that we can test the 'main effect' for each independent variable and also explore the possibility of an 'interaction effect'. An interaction effect occurs when the effect of one independent variable on the dependent variable depends on the level of a second independent variable. For example, in this case we may find that the influence of age on optimism is different for males and females. For males, optimism may increase with age, while for females it may decrease. If that were the case, we would say that there is an interaction effect. In order to describe the impact of age, we must specify which group (males/females) we are referring to.

If you are not clear on main effects and interaction effects I suggest you review this material in any good statistics text (see Reference list at the end of the chapter). Before proceeding, I would also recommend that you read through the introduction to Part Five of this book, where I discuss a range of topics relevant to analysis of variance techniques.

Details of example

To demonstrate the use of this technique I will be using the survey.sav data file included on the website accompanying this book (see p. xi). The data come from a survey that was conducted to explore the factors that affect respondents' psychological adjustment, health and wellbeing. This is a real data file from actual research conducted by a group of my Graduate Diploma students. Full details of the study, the questionnaire and scales used are provided in the Appendix. If you wish to follow along with the steps described in this chapter you should start SPSS and open the file labelled survey.sav. This data file can be opened only in SPSS.

Details of the variables used in this analysis are provided in the table below:

File name	Variable name	Variable label	Coding instructions
survey.sav	Toptim	Total optimism	Total score on the Optimism scale. Scores can range from 6 to 30 with high scores indicating higher levels of optimism.
	Agegp3	Agegp3	This variable is a recoded variable, dividing age into three equal groups (see instructions for how to do this in Chapter 8): Group 1: 18–29 = 1 Group 2: 30–44 = 2 Group 3: 45+ = 3
	sex	sex	Males=1, Females=2

Summary for two-way ANOVA

Example of research question: What is the impact of age and gender on optimism? Does gender moderate the relationship between age and optimism?

What you need: Three variables:
- two categorical independent variables (e.g. Sex: males/females; Age group: young, middle, old); and
- one continuous dependent variable (e.g. total optimism).

What it does: Two-way ANOVA allows you to simultaneously test for the effect of each of your independent variables on the dependent variable and also identifies any interaction effect.

For example, it allows you to test for:
- sex differences in optimism;
- differences in optimism for young, middle and old subjects; and
- the interaction of these two variables—is there a difference in the effect of age on optimism for males and females?

Assumptions: See the introduction to Part Five for a discussion of the assumptions underlying ANOVA.

Non-parametric alternative: none

Procedure for two-way ANOVA

1. From the menu at the top of the screen click on: **Analyze**, then click on **General Linear Model**, then on **Univariate**.

2. Click on your dependent, continuous variable (e.g. total optimism) and move it into the box labelled **Dependent variable**.

3. Click on your two independent, categorical variables (sex, agegp3: this is age grouped into three categories) and move these into the box labelled **Fixed Factors**.

4. Click on the **Options** button.
   - Click on **Descriptive Statistics**, **Estimates of effect size** and **Homogeneity tests**.
   - Click on **Continue**.

5. Click on the **Post Hoc** button.
   - From the **Factors** listed on the left-hand side choose the independent variable(s) you are interested in (this variable should have three or more levels or groups: e.g. agegp3).
   - Click on the arrow button to move it into the **Post Hoc Tests for** section.
   - Choose the test you wish to use (in this case **Tukey**).
   - Click on **Continue**.

6. Click on the **Plots** button.
   - In the **Horizontal** box put the independent variable that has the most groups (e.g. agegp3).
   - In the box labelled **Separate Lines** put the other independent variable (e.g. sex).
   - Click on **Add**.
   - In the section labelled **Plots** you should now see your two variables listed (e.g. agegp3*sex).

7. Click on **Continue** and then **OK**.

The output generated from this procedure is shown below.

Descriptive Statistics

Dependent Variable: Total Optimism

AGEGP	SEX	Mean	Std. Deviation	N
18-29	MALES	21.38	4.33	60
	FEMALES	21.34	4.72	87
	Total	21.36	4.55	147
30-44	MALES	22.38	3.55	68
	FEMALES	21.88	4.58	85
	Total	22.10	4.15	153
45+	MALES	22.23	4.09	56
	FEMALES	23.47	4.70	79
	Total	22.96	4.49	135
Total	MALES	22.01	3.98	184
	FEMALES	22.20	4.73	251
	Total	22.12	4.43	435

Levene's Test of Equality of Error Variances[a]

Dependent Variable: Total Optimism

F	df1	df2	Sig.
1.083	5	429	.369

Tests the null hypothesis that the error variance of the dependent variable is equal across groups.

[a] Design: Intercept +AGEGP3 +SEX+AGEGP3 * SEX

Tests of Between-Subjects Effects

Dependent Variable: total optimism

Source	Type III Sum of Squares	df	Mean Square	F	Sig.	Partial Eta Squared
Corrected Model	238.647[a]	5	47.729	2.475	.032	.028
Intercept	206790.069	1	206790.069	10721.408	.000	.962
agegp3	150.863	2	75.431	3.911	.021	.018
sex	5.717	1	5.717	.296	.586	.001
agegp3 * sex	55.709	2	27.855	1.444	.237	.007
Error	8274.374	429	19.288			
Total	221303.000	435				
Corrected Total	8513.021	434				

a. R Squared = .028 (Adjusted R Squared = .017)

Multiple Comparisons

Dependent Variable: Total Optimism

Tukey HSD

(I) AGEGP3	(J) AGEGP3	Mean Difference (I-J)	Std. Error	Sig.	95% Confidence Interval	
					Lower Bound	Upper Bound
18-29	30-44	-.74	.51	.307	-1.93	.44
	45+	-1.60*	.52	.007	-2.82	-.37
30-44	18-29	.74	.51	.307	-.44	1.93
	45+	-.85	.52	.228	-2.07	.36
45+	18-29	1.60*	.52	.007	.37	2.82
	30-44	.85	.52	.228	-.36	2.07

Based on observed means.

* The mean difference is significant at the .05 level.

Profile plots

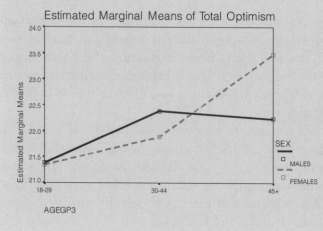

Estimated Marginal Means of Total Optimism

Interpretation of output from two-way ANOVA

The output from ANOVA gives you a number of tables and plots with useful information concerning the effect of your two independent variables. A highlighted tour of this output follows:

- **Descriptive statistics.** These provide the mean scores, standard deviations and N for each subgroup. Check that these values are correct. Inspecting the pattern of these values will also give you an indication of the impact of your independent variables.

- **Levene's Test of Equality of Error Variances.** This test provides a test of one of the assumptions underlying analysis of variance. The value you are most interested in is the Sig. level. You want this to be *greater* than .05, and therefore *not* significant. A significant result (Sig. value less than .05) suggests that the variance of your dependent variable across the groups is not equal. If you find this to be the case in your study it is recommended that you set a more stringent significance level (e.g. .01) for evaluating the results of your two-way ANOVA. That is, you will consider the main effects and interaction effects significant only if the Sig. value is greater than .01. In the example displayed above, the Sig. level is .369. As this is larger than .05, we can conclude that we have not violated the homogeneity of variances assumption.

The main output from two-way ANOVA is a table labelled **Tests Of Between-Subjects Effects.** This gives you a number of pieces of information, not necessarily in the order in which you need to check them.

Interaction effects

The first thing you need to do is to check for the possibility of an interaction effect (e.g. that the influence of age on optimism levels depends on whether you are a male or a female). If you find a significant interaction effect you cannot easily and simply interpret the main effects. This is because, in order to describe the influence of one of the independent variables, you need to specify the level of the other independent variable.

In the SPSS output the line we need to look at is labeled AGEGP3*SEX. To find out whether the interaction is significant, check the **Sig.** column for that line. If the value is less than or equal to .05 (e.g. .03, .01, .001), then there is a significant interaction effect. In our example the interaction effect is not significant (agegp3*sex: sig.=.237). This indicates that there is no significant difference in the effect of age on optimism for males and females.

Warning: When checking significance levels in this output, make sure you read the correct column (the one labelled **Sig.**—a lot of students make the mistake of reading the **Partial Eta Squared** column, with dangerous consequences!).

Main effects

We did not have a significant interaction effect; therefore, we can safely interpret the main effects. These are the simple effect of one independent variable (e.g. the effect of sex with all age groups collapsed). In the left-hand column, find the variable you are interested in (e.g. AGEGP3) To determine whether there is a main effect for each independent variable, check in the column marked **Sig.** next to each variable. If the value is less than or equal to .05 (e.g. .03, .01, .001), then there is a significant main effect for that independent variable. In the example shown above there is a significant main effect for age group (agegp3: sig=.021), but no significant main effect for sex (sex: sig=.586). This means that males and females do not differ in terms of their optimism scores, but there is a difference in scores for young, middle and old subjects.

Effect size

The effect size for the agegp3 variable is provided in the column labelled **Partial Eta Squared** (.018). Using Cohen's (1988) criterion, this can be classified as small (see introduction to Part Five). So, although this effect reaches statistical significance, the actual difference in the mean values is very small. From the **Descriptives** table we can see that the mean scores for the three age groups (collapsed for sex) are 21.36, 22.10, 22.96. The difference between the groups appears to be of little practical significance.

Post-hoc tests

Although we know that our age groups differ, we do not know where these differences occur: Is gp1 different from gp2, is gp2 different from gp3, is gp1 different from gp3? To investigate these questions we need to conduct post-hoc tests (see description of these in the introduction to Part Five). Post-hoc tests are relevant only if you have more than two levels (groups) to your independent variable. These tests systematically compare each of your pairs of groups, and indicate whether there is a significant difference in the means of each. SPSS provides these post-hoc tests as part of the ANOVA output. You are, however, not supposed to look at them until you find a significant main effect or interaction effect in the overall (omnibus) analysis of variance test. In this example we obtained a significant main effect for agegp3 in our ANOVA; therefore, we are entitled to dig further using the post-hoc tests for agegp.

Multiple comparisons

The results of the post-hoc tests are provided in the table labelled **Multiple Comparisons**. We have requested the Tukey Honestly Significant Difference test, as this is one of the more commonly used tests. Look down the column labelled **Sig.** for any values less than .05. Significant results are also indicated by a little asterisk in the column labelled **Mean Difference**. In the above example only group 1 (18–29) and group 3 (45+) differ significantly from one another.

Plots

You will see at the end of your SPSS output a plot of the optimism scores for males and females, across the three age groups. This plot is very useful for allowing you to visually inspect the relationship among your variables. This is often easier than trying to decipher a large table of numbers. Although presented last, the plots are often useful to inspect first to help you better understand the impact of your two independent variables.

Warning: When interpreting these plots, remember to consider the scale used to plot your dependent variable. Sometimes what looks like an enormous difference on the plot will involve only a few points difference. You will see this in the current example. In the first plot there appears to be quite a large difference in male and female scores for the older age group (45+). If you read across to the scale, however, the difference is only small (22.2 as compared with 23.5).

Presenting the results from two-way ANOVA

Strange looking numbers

In your output you might come across strange looking numbers which take the form: 1.24E–02. These numbers have been written in scientific notation. They can easily be converted to a 'normal' number for presentation in your results. The number after the E tells you how many places you need to shift the decimal point. If the number after the E is negative, shift the decimal point to the left; if it is positive, shift it to the right. Therefore, 1.24E–02 becomes 0.124. The number 1.24E02 would be written as 124.

You can prevent these small numbers from being displayed in this strange way. Go to **Edit, Options** and put a tick in the box labelled: **No scientific notation for small numbers in tables**. Rerun the analysis and normal numbers should appear in all your output.

The results of the analysis conducted above could be presented as follows:

A two-way between-groups analysis of variance was conducted to explore the impact of sex and age on levels of optimism, as measured by the Life Orientation test (LOT). Subjects were divided into three groups according to their age (Group 1: 18–29 years; Group 2: 30–44 years; Group 3: 45 years and above). There was a statistically significant main effect for age [$F_{(2, 429)}=3.91$, $p=.02$]; however, the effect size was small (partial eta squared=.02). Post-hoc comparisons using the Tukey HSD test indicated that the mean score for the 18–29 age group ($M=21.36$, $SD=4.55$) was significantly different from the 45 + group ($M=22.96$, $SD=4.49$). The 30–44 age group ($M=22.10$, $SD=4.15$) did not differ significantly from either of the other groups. The main effect for sex [$F_{(1, 429)}=.30$, $p=.59$] and the interaction effect [$F_{(2, 429)}=1.44$, $p=.24$] did not reach statistical significance.

You will notice that in the results above I reported the main effects first, before discussing the interaction effect (despite earlier telling you to check interaction effects first). This was because the interaction effect was not significant in this example. If my interaction effect had been significant I would have reported that first, because it influences how we interpret the effect of each independent variable.

Additional analyses if you obtain a significant interaction effect

If you obtain a significant result for your interaction effect you may wish to conduct follow-up tests to explore this relationship further (this applies only if one of your variables has three or more levels). One way that you can do this is to conduct an analysis of simple effects. This means that you will look at the results for each of the subgroups separately. This involves splitting the sample into groups according to one of your independent variables and running separate one-way ANOVAs to explore the effect of the other variable. If we had obtained a significant interaction effect in the above example we might choose to split the file by sex and look at the effect of age on optimism separately for males and females.

To split the sample and repeat analyses for each group you need to use the SPSS **Split File** option. This option allows you to split your sample according to one categorical variable and to repeat analyses separately for each group.

Procedure for splitting the sample

1. From the menu at the top of the screen click on: **Data**, then click on **Split File**.

2. Click on **Organize output by groups**.

3. Move the grouping variable (sex) into the box marked **Groups based on**.

4. This will split the sample by sex and repeat any analyses that follow for these two groups separately.

5. Click on **OK**.

After splitting the file you then perform a one-way ANOVA (see Chapter 17), comparing optimism levels for the three age groups. With the **Split File** in operation you will obtain separate results for males and females.

Important

When you have finished these analyses for the separate groups you must turn the **Split File** option off if you want to do further analyses using the whole sample. To do this, go to the **Data** menu, click on **Split File**, and then click on the first circle: **Analyze all cases, do not create groups**. Then click on **OK**.

Additional exercises

Business

Data file: *staffsurvey.sav*. See Appendix for details of the data file.

1. Conduct a two-way ANOVA with post-hoc tests (if appropriate) to compare staff satisfaction scores (*totsatis*) across each of the length of service categories (use the *servicegp3* variable) for permanent versus casual staff (*employstatus*).

Health

Data file: *sleep.sav*. See Appendix for details of the data file.

1. Conduct a two-way ANOVA with post-hoc tests (if appropriate) to compare male and female (*gender*) mean sleepiness ratings (Sleepiness and Associated Sensations Scale total score: *totSAS*) for the three age groups defined by the variable *agegp3* (<=37, 38–50, 51+).

References

Cohen, J. (1988). *Statistical power analysis for the behavioral sciences*. Hillsdale, NJ: Erlbaum.

Gravetter, F. J., & Wallnau, L. B. (2000). *Statistics for the behavioral sciences* (5th edn). Belmont, CA: Wadsworth.

19 Mixed between-within subjects analysis of variance

In the previous analysis of variance chapters we have explored the use of both between-subjects designs (comparing two or more different groups) and within-subjects or repeated measures designs (one group of subjects exposed to two or more conditions). Up until now, we have treated these approaches separately. There may be situations, however, where you want to combine the two approaches in the one study, with one independent variable being between-subjects and the other a within-subjects variable. For example, you may want to investigate the impact of an intervention on clients' anxiety levels (using pre-test and post-test), but you would also like to know whether the impact is different for males and females. In this case you have two independent variables: one is a between-subjects variable (gender: males/females), the other is a within-subjects variable (time). In this case you would expose a group of both males and females to the intervention and measure their anxiety levels at Time 1 (pre-intervention) and again at Time 2 (after the intervention).

SPSS allows you to combine between-subjects and within-subjects variables in the one analysis. You will see this analysis referred to in some texts as a split-plot ANOVA design (SPANOVA). I have chosen to use Tabachnick and Fidell's (2001) term 'mixed between-within subjects ANOVA' because I feel this best describes what is involved. This technique is an extension to the repeated measures design discussed previously in Chapter 17. It would be a good idea to review that chapter before proceeding further.

This chapter is intended as a very brief overview of mixed between-within subjects ANOVA. For a more detailed discussion of this technique, see some of the additional readings suggested in the References section in the introduction to Part Five. Post-hoc tests associated with mixed designs are not discussed here; for advice on this topic, see Stevens (1996, Chapter 13).

Details of example

To illustrate the use of mixed between-within subjects ANOVA I will be using the experim.sav data file included on the website that accompanies this book (see p. xi). These data refer to a fictitious study that involves testing the impact of two different types of interventions in helping students cope with their anxiety concerning a forthcoming statistics course (see the Appendix for full details of the study). Students were divided into two equal groups and asked to complete a Fear of Statistics test. One group was given a number of sessions designed to improve their mathematical skills, the second group participated in a program designed to build their confidence. After the program they were again asked to complete the same scale they'd done before the program. They were also followed up three months later. The manufactured data file is included on the website that accompanies this book. If you wish to follow the procedures detailed below you will need to start SPSS and open the data file labelled experim.sav. This file can be opened only in SPSS.

In this example I will compare the impact of the Maths Skills class (Group 1) and the Confidence Building class (Group 2) on participants' scores on the Fear of Statistics test, across the three time periods. Details of the variables names and labels from the data file are provided in the following table.

File name	Variable name	Variable label	Coding instructions
experim.sav	Group	Type of class	1=Maths Skills 2=Confidence Building
	Fost1	Fear of Statistics scores at time1	Total scores on the Fear of Statistics test administered prior to the program. Scores range from 20 to 60. High scores indicate greater fear of statistics.
	Fost2	Fear of Statistics scores at time2	Total scores on the Fear of Statistics test administered after the program was complete. Scores range from 20 to 60. High scores indicate greater fear of statistics.
	Fost3	Fear of Statistics scores at time3	Total scores on the Fear of Statistics test administered 3 months after the program was complete. Scores range from 20 to 60. High scores indicate a greater fear of statistics.

Summary for mixed between-within ANOVA

Example of research question: Which intervention is more effective in reducing participants' Fear of Statistics test scores across the three time periods (pre-intervention, post-intervention, follow-up)?

Is there a change in participants' fear of statistics scores across the three time periods (before the intervention, after the intervention and three months later)?

What you need: At least three variables are involved:
- one categorical independent between-subjects variable with two or more levels (group1/group2);

- one categorical independent within-subjects variable with two or more levels (time1/time2/ time3); and
- one continuous dependent variable (scores on the Fear of Statistics test measured at each time period).

What it does:

This analysis will test whether there are main effects for each of the independent variables and whether the interaction between the two variables is significant.

In this example it will tell us whether there is a change in fear of statistics scores over the three time periods (main effect for time). It will compare the two interventions (maths skills/confidence building) in terms of their effectiveness in reducing fear of statistics (main effect for group). Finally, it will tell us whether the change in fear of statistics scores over time is different for the two groups (interaction effect).

Assumptions:

See the introduction to Part Five for a discussion of the general assumptions underlying ANOVA.

Additional assumption:

Homogeneity of intercorrelations. For each of the levels of the between-subjects variable the pattern of inter-correlations among the levels of the within-subjects variable should be the same. This assumption is tested as part of the analysis, using Box's M statistic. Because this statistic is very sensitive, a more conservative alpha level of .001 should be used. You are hoping that the statistic is not significant (i.e. the probability level should be greater than .001).

Non-parametric alternative:

Procedure for mixed between-within ANOVA

1. From the menu at the top of the screen click on: **Analyze**, then click on **General Linear Model**, then on **Repeated measures**.

2. In the box labelled **Within-Subject Factor Name** type a name that describes the within-subjects factor (e.g. time). This is not an actual variable name, but a descriptive term that you choose.

3. In the **Number of Levels** box type the number of levels that this factor has (in this case there are three time periods; therefore, you would type 3).

4. Click on the **Add** button. Click on the **Define** button.

5. Click on the variables that represent the within-subjects factor (e.g. fear of stats scores from Time 1, Time 2 and Time 3).

6. Click on the arrow to move these into the **Within-Subjects Variables** box. You will see them listed (using only the short variable name: fost1, fost2, fost3).

7. Click on your between-subjects variable (e.g. type of class). Click on the arrow to move this variable into the **Between-Subjects Factors** box.

8. Click on the **Options** button.
   - In the **Display** section click on **Descriptive statistics, Estimates of effect size, Homogeneity tests**.
   - Click on **Continue**.

9. Click on the **Plots** button.
   - Click on the within-groups factor (e.g. time) and move it into the box labelled **Horizontal Axis**.
   - In the **Separate Lines** box click on the between-groups variable (e.g. group).

10. Click on **Add**. In the box, you should see your variables listed (e.g. time*group).

11. Click on **Continue** and then **OK**.

The output generated from this procedure is shown below. Only a selected sample of the output is displayed.

Descriptive Statistics

	Type of class	Mean	Std. Deviation	N
fear of stats time1	maths skills	39.87	4.60	15
	confidence building	40.47	5.82	15
	Total	40.17	5.16	30
fear of stats time2	maths skills	37.67	4.51	15
	confidence building	37.33	5.88	15
	Total	37.50	5.15	30
fear of stats time3	maths skills	36.07	5.43	15
	confidence building	34.40	6.63	15
	Total	35.23	6.02	30

Box's Test of Equality of Covariance Matrices [a]

Box's M	1.520
F	.224
df1	6
df2	5680
Sig.	.969

Tests the null hypothesis that the observed covariance matrices of the dependent variables are equal across groups.

[a] Design: Intercept +GROUP
Within Subjects Design: TIME

Multivariate Tests[b]

Effect		Value	F	Hypothesis df	Error df	Sig.	Partial Eta Squared
time	Pillai's Trace	.663	26.593[a]	2.000	27.000	.000	.663
	Wilks' Lambda	.337	26.593[a]	2.000	27.000	.000	.663
	Hotelling's Trace	1.970	26.593[a]	2.000	27.000	.000	.663
	Roy's Largest Root	1.970	26.593[a]	2.000	27.000	.000	.663
time * group	Pillai's Trace	.131	2.034[a]	2.000	27.000	.150	.131
	Wilks' Lambda	.869	2.034[a]	2.000	27.000	.150	.131
	Hotelling's Trace	.151	2.034[a]	2.000	27.000	.150	.131
	Roy's Largest Root	.151	2.034[a]	2.000	27.000	.150	.131

a. Exact statistic

b.
Design: Intercept+group
Within Subjects Design: time

Mauchly's Test of Sphericity[b]

Measure: MEASURE_1

Within Subjects Effect	Mauchly's W	Approx. Chi-Square	df	Sig.	Epsilon[a]		
					Greenhouse-Geisser	Huynh-Feldt	Lower-bound
TIME	.348	28.517	2	.000	.605	.640	.500

Tests the null hypothesis that the error covariance matrix of the orthonormalized transformed dependent variables is proportional to an identity matrix.

a. May be used to adjust the degrees of freedom for the averaged tests of significance. Corrected tests are displayed in the Tests of Within-Subjects Effects table.

b. Design: Intercept +GROUP
Within Subjects Design: TIME

Levene's Test of Equality of Error Variances[a]

	F	df1	df2	Sig.
fear of stats time1	.893	1	28	.353
fear of stats time2	.767	1	28	.389
fear of stats time3	.770	1	28	.388

Tests the null hypothesis that the error variance of the dependent variable is equal across groups.

a. Design: Intercept +GROUP
Within Subjects Design: TIME

Tests of Between-Subjects Effects

Measure: MEASURE_1
Transformed Variable: Average

Source	Type III Sum of Squares	df	Mean Square	F	Sig.	Partial Eta Squared
Intercept	127464.100	1	127464.100	1531.757	.000	.982
group	4.900	1	4.900	.059	.810	.002
Error	2330.000	28	83.214			

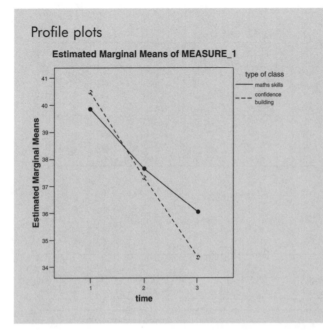

Profile plots

Interpretation of output from mixed between-within ANOVA

You will notice (once again) that this SPSS technique generates quite a good deal of rather complex-looking output. If you have worked your way through the previous chapters you will recognise some of the output from other analysis of variance procedures. This output provides tests for the assumptions of sphericity, univariate ANOVA results and also multivariate ANOVA results. Full discussion of the difference between the univariate and multivariate results is beyond the scope of this book; in this chapter only the multivariate results will be discussed (for more information see Stevens, 1996, pp. 466–469). The reason for interpreting the multivariate statistics provided by SPSS is that the univariate statistics make the assumption of sphericity. The sphericity assumption requires that the variance of the population difference scores for any two conditions are the same as the variance of the population difference scores for any other two conditions (an assumption that is commonly violated). This is assessed by SPSS using Mauchly's Test of Sphericity.

The multivariate statistics do not require sphericity. You will see in our example that we have violated the assumption of sphericity, as indicated by the Sig. value of .000 in the box labelled **Mauchly's Test of Sphericity**. Although there are ways to compensate for this assumption violation, it is safer to inspect the multivariate statistics provided in the output.

Let's look at the key values in the output that you will need to consider.

Descriptive statistics

In the first output box you are provided with the descriptive statistics for your three sets of scores (Mean, Standard deviation, N). It is a good idea to check that these make sense. Are there the right numbers of people in each group? Do the mean values make sense given the scale that was used? In the example above you will see that the highest fear of statistics scores are at Time 1 (39.87 and 40.47), that they drop at Time 2 (37.67 and 37.33) and drop even further at Time 3 (36.07 and 34.40). What we don't know, however, is whether these differences are large enough to be considered statistically significant.

Interaction effect

Before we can look at the main effects we need first to assess the interaction effect. Is there the same change in scores over time for the two different groups (maths skills/confidence building)? This is indicated in the second set of rows in the **Multivariate Tests** table (TIME*GROUP). The value that you are interested in is **Wilks' Lambda** and the associated probability value given in the column labelled Sig. All of the multivariate tests yield the same result; however, the most commonly reported statistic is Wilks' Lambda. In this case the interaction effect is not statistically significant (the Sig. level for Wilks' Lambda is .15, which is greater than our alpha level of .05).

Main effects

We can now move on and assess the main effects for each of our independent variables. In this example the value for Wilks' Lambda for time is .337, with a probability value of .000 (which really means p<.0005). Because our p value is less than .05, we can conclude that there is a statistically significant effect for time. This suggests that there was a change in fear of statistics scores across the three different time periods. The main effect for time was significant.

Although we have found a statistically significant difference among the time periods, we also need to assess the effect size of this result (see discussion on effect sizes in the introduction to Part Five). The value you are interested in is **Partial Eta Squared**, given in the **Multivariate Tests** output box. The value obtained for time in this study is .663. Using the commonly used guidelines proposed by Cohen (1988) (.01=small effect, .06=moderate effect, .14=large effect), this result suggests a very large effect size.

Now that we have explored the within-subjects effects, we need to consider the main effect of our between-subjects variable (type of class: maths skills/confidence building).

Strange looking numbers

In your output you might come across strange looking numbers which take the form: 1.24E–02. These numbers have been written in scientific notation. They can easily be converted to a 'normal' number for presentation in your results. The number after the E tells you how many places you need to shift the decimal point. If the number after the E is negative, shift the decimal point to the left; if it is positive, shift it to the right. Therefore, 1.24E–02 becomes 0.124. The number 1.24E02 would be written as 124.

You can prevent these small numbers from being displayed in this strange way. Go to **Edit**, **Options** and put a tick in the box labelled: **No scientific notation for small numbers in tables**. Rerun the analysis and normal numbers should appear in all your output.

Between-subjects effect

The results that we need to look at are in the table labelled **Tests of Between-Subjects Effects**. Read across the row labelled Group (this is the shortened SPSS variable name for the type of class). The Sig. value is .81. This is not less than our alpha level of .05, so we conclude that the main effect for group is not significant. There was no significant difference in the fear of statistics scores for the two groups (those who received maths skills training and those who received the confidence building intervention).

The effect size of the between-subject effect is also given in the **Tests of Between-Subject Effects** table. The partial eta-squared value for Group in this case is .002. This is very small. It is therefore not surprising that it did not reach statistical significance.

Presenting the results from mixed between-within ANOVA

The method of presenting the results for this technique is a combination of that used for a between-groups ANOVA (see Chapter 18) and a repeated measures ANOVA (see Chapter 17). Report the interaction effect and main effects for each independent variable and associated effect sizes.

References

Cohen, J. (1988). *Statistical power analysis for the behavioral sciences*. Hillsdale, NJ: Erlbaum.

Stevens, J. (1996). *Applied multivariate statistics for the social sciences* (3rd edn). Mahway, NJ: Lawrence Erlbaum.

Tabachnick, B. G., & Fidell, L. S. (2001). *Using multivariate statistics* (4th edn). New York: HarperCollins.

20 Multivariate analysis of variance

In previous chapters we explored the use of analysis of variance to compare groups on a *single* dependent variable. In many research situations, however, we are interested in comparing groups on a range of different characteristics. This is quite common in clinical research, where the focus is on the evaluation of the impact of an intervention on a variety of outcome measures (e.g. anxiety, depression, physical symptoms).

Multivariate analysis of variance (MANOVA) is an extension of analysis of variance for use when you have *more than one* dependent variable. These dependent variables should be related in some way, or there should be some conceptual reason for considering them together. MANOVA compares the groups and tells you whether the mean differences between the groups on the combination of dependent variables is likely to have occurred by chance. To do this MANOVA creates a new summary dependent variable, which is a linear combination of each of your original dependent variables. It then performs an analysis of variance using this new combined dependent variable. MANOVA will tell you if there is a significant difference between your groups on this composite dependent variable; it also provides the univariate results for each of your dependent variables separately.

Some of you might be thinking: Why not just conduct a series of ANOVAs separately for each dependent variable? This is in fact what many researchers do. Unfortunately by conducting a whole series of analyses you run the risk of an 'inflated Type 1 error'. (See the introduction to Part Five for a discussion of Type 1 and Type 2 errors.) Put simply, this means that the more analyses you run, the more likely you are to find a significant result, even if in reality there are no differences between your groups. The advantage of using MANOVA is that it 'controls' or adjusts for this increased risk of a Type 1 error; however, this comes at a cost. MANOVA is a much more complex set of procedures, and it has a number of additional assumptions that must be met.

If you decide that MANOVA is a bit out of your depth just yet, all is not lost. If you have a number of dependent variables you can still perform a series of ANOVAs separately for each dependent variable. If you choose to do this, you might like to reduce the risk of a Type 1 error by setting a more stringent alpha value. One way to control for the Type 1 error across multiple tests is to use a Bonferroni adjustment. To do this you divide your normal alpha value (typically .05) by the number of tests that you are intending to perform. If there are three dependent variables, you would divide .05 by 3 (which equals .017 after rounding) and you would use this new value as your cut-off. Differences between your groups would need a probability value of less than .017 before you could consider them statistically significant.

MANOVA can be used in one-way, two-way and higher-order factorial designs (with multiple independent variables), and when using analysis of covariance (controlling for an additional variable). In the example provided in this chapter a simple one-way MANOVA is demonstrated. Coverage of more complex designs is beyond the scope of this book; however, for students interested in the use of MANOVA with two-way and higher designs, a number of suggested readings are provided in the References at the end of the chapter.

Details of example

To demonstrate MANOVA I have used the data file survey.sav on the website that accompanies this book (see p. xi). For a full description of this study please see the Appendix. In this example the difference between males and females on a number of measures of wellbeing is explored. These include a measure of negative mood (Negative Affect scale), positive mood (Positive Affect scale) and perceived stress (Total Perceived Stress scale). If you wish to follow along with the procedures described below you should start SPSS and open the file labelled survey.sav. This file can be opened only in SPSS.

Summary for one-way MANOVA

Example of research question:	Do males and females differ in terms of overall wellbeing? Are males better adjusted than females in terms of their positive and negative mood states and levels of perceived stress?
What you need:	One-way MANOVA: • one categorical, independent variable (e.g. sex); and • two or more continuous, dependent variables (e.g. negative affect, positive affect, perceived stress). MANOVA can also be extended to two-way and higher-order designs involving two or more categorical, independent variables.
What it does:	Compares two or more groups in terms of their means on a group of dependent variables. Tests the null hypothesis that the population means on a set of dependent variables do not vary across different levels of a factor or grouping variable.
Assumptions:	MANOVA has a number of assumptions. These are discussed in more detail in the next section. You should

also review the material on assumptions in the
introduction to Part Five of this book.

1. Sample size
2. Normality
3. Outliers
4. Linearity
5. Homogeneity of regression
6. Multicollinearity and singularity
7. Homogeneity of variance-covariance matrices

Non-parametric alternative: none

Assumption testing

Before proceeding with the main MANOVA analysis we will test whether our
data conform to the assumptions listed in the Summary. Some of these tests are
not strictly necessary given our large sample size, but I will demonstrate them
so that you can see the steps involved.

1. Sample size

You need to have more cases in each cell than you have dependent variables.
Ideally you will have more than this, but this is the absolute minimum. Having
a larger sample can also help you 'get away with' violations of some of the
other assumptions (e.g. normality). The minimum required number of cases in
each cell in this example is three (the number of dependent variables). We have
a total of six cells (two levels of our independent variable: male/female; and
three dependent variables for each). The number of cases in each cell is provided
as part of the MANOVA output. In our case we have many more than the
required number of cases per cell (see the **Descriptive statistics** box in the
Output).

2. Normality

Although the significance tests of MANOVA are based on the multivariate normal
distribution, in practice it is reasonably robust to modest violations of normality
(except where the violations are due to outliers). According to Tabachnick and
Fidell (2001, p. 329), a sample size of at least 20 in each cell should ensure
'robustness'. You need to check both (a) univariate normality (see Chapter 6)
and (b) multivariate normality (using something called Mahalanobis distances).
The procedures used to check for normality will also help you identify any outliers
(see Assumption 3).

3. Outliers

MANOVA is quite sensitive to outliers (i.e. data points or scores that are different from the remainder of the scores). You need to check for univariate outliers (for each of the dependent variables separately) and multivariate outliers. Multivariate outliers are subjects with a strange combination of scores on the various dependent variables (e.g. very high on one variable, but very low on another). Check for univariate outliers by using **Explore** (see Chapter 6). The procedure to check for multivariate outliers and multivariate normality is demonstrated below.

Checking multivariate normality

To test for multivariate normality we will be asking SPSS to calculate Mahalanobis distances using the **Regression** menu. Mahalanobis distance is the distance of a particular case from the centroid of the remaining cases, where the centroid is the point created by the means of all the variables (Tabachnick & Fidell, 2001, p. 68). This analysis will pick up on any cases that have a strange pattern of scores *across* the three dependent variables.

The procedure detailed below will create a new variable in your data file (labelled mah_1). Each person or subject receives a value on this variable which indicates the degree to which their pattern of scores differ from the remainder of the sample. To decide whether a case is an outlier you need to compare the Mahalanobis distance value against a critical value (this is obtained using a chi-square critical value table). If an individual's mah_1 score exceeds this value, it is considered an outlier. MANOVA can tolerate a few outliers, particularly if their scores are not too extreme and you have a reasonable size data file. With too many outliers, or very extreme scores, you may need to consider deleting the cases, or alternatively transforming the variables involved (see Tabachnick & Fidell, 2001, pp. 80–82).

Procedure for obtaining Mahalanobis distances

1. From the menu at the top of the screen click on: **Analyze**, then click on **Regression**, then on **Linear**.

2. Click on the variable in your data file which uniquely identifies each of your cases (in this case it is ID). Move this variable into the **Dependent** box.

3. In the **Independent** box, put the continuous *dependent* variables that you will be using in your MANOVA analysis (e.g. total negative affect, total positive affect, total perceived stress).

4. Click on the **Save** button. In the section marked **Distances**, click on **Mahalanobis**.

5. Click on **Continue** and then **OK**.

The output generated from this procedure is shown below.

Residuals Statistics[a]

	Minimum	Maximum	Mean	Std. Deviation	N
Predicted Value	239.04	317.73	276.25	13.58	432
Std. Predicted Value	-2.741	3.056	.000	1.000	432
Standard Error of Predicted Value	8.34	34.95	15.22	4.87	432
Adjusted Predicted Value	233.16	316.81	276.23	13.65	432
Residual	-297.10	448.06	-6.47E-14	165.47	432
Std. Residual	-1.789	2.698	.000	.997	432
Stud. Residual	-1.805	2.711	.000	1.001	432
Deleted Residual	-302.42	452.38	1.25E-02	167.02	432
Stud. Deleted Residual	-1.810	2.732	.000	1.002	432
Mahal. Distance	.090	18.101	2.993	2.793	432
Cook's Distance	.000	.025	.002	.003	432
Centered Leverage Value	.000	.042	.007	.006	432

a. Dependent Variable: ID

Towards the bottom of this table you will see a row labelled **Mahal. distance**. Look across under the column marked Maximum. Take note of this value (in this example it is 18.1). You will be comparing this number to a critical value.

This critical value is determined by using a critical values of chi-square table, with the number of dependent variables that you have as your degrees of freedom (df) value. The alpha value that you use is .001. To simplify this whole process I have summarised the key values for you in Table 20.1, for studies up to ten dependent variables.

Find the column with the number of dependent variables that you have in your study (in this example it is 3). Read across to find the critical value (in this case it is 16.27).

Number of dependent variables	Critical value	Number of dependent variables	Critical value	Number of dependent variables	Critical value
2	13.82	5	20.52	8	26.13
3	16.27	6	22.46	9	27.88
4	18.47	7	24.32	10	29.59

Table 20.1

Critical values for evaluating Mahalanobis distance values

Source: Extracted and adapted from a table in Tabachnik and Fidell (1996); originally from Pearson, E. S. and Hartley, H. O. (Eds) (1958). *Biometrika tables for statisticians* (vol. 1, 2nd edn). New York: Cambridge University Press.

Compare the maximum value you obtained from your output (e.g. 18.1) with this critical value. If your value is larger than the critical value, you have 'multivariate outliers' in your data file. In my example, at least one of my cases exceeded the critical value of 16.27, suggesting the presence of multivariate outliers. I will need to do further investigation to find out how many cases are involved and just how different they are from the remaining cases. If the maximum value for Mahal. distance was less than the critical value, I could safely have assumed that there were no substantial multivariate outliers and proceeded to check other assumptions.

To find out more about these outliers I need to do some further analyses using the **Descriptive statistics** menu of SPSS.

Procedure to identify multivariate outliners

1. From the menu at the topof the screen click on: **Analyze**, then click on **Descriptive Statistics**, then on **Explore**.

2. Click on the new variable Mahalanobis Distance (mah_1), which appears at the end of your data file. Move this into the **Dependents** box.

3. Click on your identifying variable (e.g. ID), and move it into the box labelled **Label Cases by**.

4. Click on the **Statistics** button. Tick **Descriptives** and **Outliners**.

5. Click on **Continue** and then **OK**.

The output generated from this procedure is shown below.

Extreme Values

			Case Number	ID	Value
Mahalanobis Distance	Highest	1	322	415	18.10064
		2	9	9	14.48201
		3	332	425	14.21429
		4	245	307	14.16941
		5	347	440	13.70297
	Lowest	1	315	408	.09048
		2	79	90	.09048
		3	160	202	.12461
		4	417	527	.12770
		5	10	10	.12952

The **Extreme Values** box gives you the highest and lowest values for the Mahalanobis distance variable and the ID number of the person that recorded these scores. Remember, we are looking for people (or cases) who had values higher than our critical value of 16.27. In our example only one person had a score that exceeded the critical value. This was the person with ID=415 and a score of 18.10. Because we only have one person and their score is not too high, I will leave this person in the data file. If there had been a lot of outlying cases, I might have needed to consider transforming this group of variables (see Chapter 8), or removing the cases from the data file. If you find yourself in this situation, make sure you read in more detail about these options before deciding to take either of these courses of action (Tabachnick & Fidell, 2001, pp. 66–77).

4. Linearity

This assumption refers to the presence of a straight-line relationship between each pair of your dependent variables. This can be assessed in a number of ways, the most straightforward of which is to generate scatterplots between each pair of your variables. We will do this separately for our groups (males and females). This involves splitting the file by sex, and then generating scatterplots. These two steps are illustrated below.

Procedure to split the file

First, you need to split your sample into groups (by your independent variable, e.g. sex). This will display the results of your scatterplots separately for each of your groups.

1. Make sure you have the **Data Editor** window open on the screen. If you have the **Viewer** window on top you will need to click on **Window** on the menu bar, and then click on the **SPSS Data Editor**.

2. From the menu at the top of the screen click on: **Data**, then click on **Split File**.

3. Click on **Organize output by groups**.

4. Click on your independent categorical variable that will be used to create the groups (e.g. sex).

5. Move this variable into the box labelled: **Groups based on**. Click on **OK**.

Procedure to generate scatterplots

Having split your file, you can now request your scatterplots. You will need to repeat this step to cover all possible pairs of your dependent variables.

1. From the menu at the top of the screen click on: **Graphs**, then click on **Scatter**.

2. Click on **Simple**. Click on the **Define** button.

3. Click on one of your dependent variables and move it into the box labelled **Y axis**.

4. Click on another of your dependent variables and move it into the box labelled **X axis**. (It does not matter in which order you enter your variables.) Click on **OK**.

5. *Remember to go back and turn your Split File option off when you have finished (see procedure below):* Click on **Data**, then click on **Split File**.

6. Click on **Analyze all cases, do not create groups**. Click on **OK**.

The output generated from this procedure is shown below.

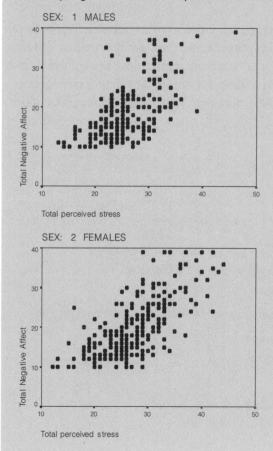

These plots do not show any evidence of non-linearity; therefore, our assumption of linearity is satisfied.

5. Homogeneity of regression

This assumption is important only if you are intending to perform a stepdown analysis. This approach is used when you have some theoretical or conceptual reason for ordering your dependent variables. It is quite a complex procedure and is beyond the scope of this book. If you are interested in finding out more, see Tabachnick and Fidell (2001, Chapter 9).

6. Multicollinearity and singularity

MANOVA works best when the dependent variables are only moderately correlated. With low correlations you should consider running separate univariate analysis of variance for your various dependent variables. When the dependent variables are highly correlated this is referred to as multicollinearity. This can occur when one of your variables is a combination of other variables (e.g. the total scores of a scale that is made up of subscales that are also included as dependent variables). This is referred to as singularity, and can be avoided by knowing what your variables are, and how the scores are obtained.

While there are quite sophisticated ways of checking for multicollinearity, the simplest way is to run **Correlation** and to check the strength of the correlations among your dependent variables (see Chapter 11). Correlations up around .8 or .9 are reason for concern. If you find any of these you may need to consider removing one of the strongly correlated pairs of dependent variables, or alternatively combining them to form a single measure.

7. Homogeneity of variance-covariance matrices

Fortunately, the test of this assumption is generated as part of your MANOVA output. The test used to assess this is Box's M Test of Equality of Covariance Matrices. This is discussed in more detail in the interpretation of the output presented below.

Performing MANOVA

The procedure for performing a one-way multivariate analysis of variance to explore sex differences in our set of dependent variables (Total Negative Affect, Total Positive Affect, Total Perceived Stress) is described below. This technique can be extended to perform a two-way or higher-order factorial MANOVA by adding additional independent variables. It can also be extended to include covariates (see analysis of covariance discussed in Chapter 21).

Procedure for MANOVA

1. From the menu at the top of the screen click on: **Analyze**, then click on **General Linear Model**, then on **Multivariate**.

2. In the **Dependent Variables** box enter each of your dependent variables (e.g. Total negative affect, Total positive Affect, Total perceived stress).

3. In the **Fixed Factors** box enter your independent variable (e.g. sex).

4. Click on the **Model** button. Make sure that the **Full factorial** button is selected in the **Specify Model** box.

5. Down the bottom in the **Sum of squares** box, **Type III** should be displayed. This is the default method of calculating sums of squares. Click on **Continue**.

6. Click on the **Options** button. In the section labelled **Factor and Factor interactions** click on your independent variable (e.g. sex). Move it into the box marked **Display Means for:**.

7. In the **Display** section of this screen, put a tick in the boxes labelled **Descriptive Statistics**, **Estimates of effect size** and **Homogeneity tests**.

8. Click on **Continue** and then **OK**.

The output generated from this procedure is shown below.

Descriptive Statistics

	SEX	Mean	Std. Deviation	N
Total Positive Affect	MALES	33.63	6.99	184
	FEMALES	33.69	7.44	248
	Total	33.66	7.24	432
Total Negative Affect	MALES	18.71	6.90	184
	FEMALES	19.98	7.18	248
	Total	19.44	7.08	432
Total perceived stress	MALES	25.79	5.41	184
	FEMALES	27.42	6.08	248
	Total	26.72	5.85	432

Box's Test of Equality of Covariance Matrices[a]

Box's M	6.942
F	1.148
df1	6
df2	1074772
Sig.	.331

Tests the null hypothesis that the observed covariance matrices of the dependent variables are equal across groups.

a. Design: Intercept+SEX

Multivariate Tests[b]

Effect		Value	F	Hypothesis df	Error df	Sig.	Partial Eta Squared
Intercept	Pillai's Trace	.987	10841.625[a]	3.000	428.000	.000	.987
	Wilks' Lambda	.013	10841.625[a]	3.000	428.000	.000	.987
	Hotelling's Trace	75.993	10841.625[a]	3.000	428.000	.000	.987
	Roy's Largest Root	75.993	10841.625[a]	3.000	428.000	.000	.987
sex	Pillai's Trace	.024	3.569[a]	3.000	428.000	.014	.024
	Wilks' Lambda	.976	3.569[a]	3.000	428.000	.014	.024
	Hotelling's Trace	.025	3.569[a]	3.000	428.000	.014	.024
	Roy's Largest Root	.025	3.569[a]	3.000	428.000	.014	.024

a. Exact statistic

b. Design: Intercept+sex

Levene's Test of Equality of Error Variances[a]

	F	df1	df2	Sig.
Total Positive Affect	1.065	1	430	.303
Total Negative Affect	1.251	1	430	.264
Total perceived stress	2.074	1	430	.151

Tests the null hypothesis that the error variance of the dependent variable is equal across groups.

a. Design: Intercept+SEX

Tests of Between-Subjects Effects

Source	Dependent Variable	Type III Sum of Squares	df	Mean Square	F	Sig.	Partial Eta Squared
Corrected Model	total positive affect	.440[a]	1	.440	.008	.927	.000
	total negative affect	172.348[b]	1	172.348	3.456	.064	.008
	total perceived stress	281.099[c]	1	281.099	8.342	.004	.019
Intercept	total positive affect	478633.634	1	478633.634	9108.270	.000	.955
	total negative affect	158121.903	1	158121.903	3170.979	.000	.881
	total perceived stress	299040.358	1	299040.358	8874.752	.000	.954
sex	total positive affect	.440	1	.440	.008	.927	.000
	total negative affect	172.348	1	172.348	3.456	.064	.008
	total perceived stress	281.099	1	281.099	8.342	.004	.019
Error	total positive affect	22596.218	430	52.549			
	total negative affect	21442.088	430	49.865			
	total perceived stress	14489.121	430	33.696			
Total	total positive affect	512110.000	432				
	total negative affect	184870.000	432				
	total perceived stress	323305.000	432				
Corrected Total	total positive affect	22596.657	431				
	total negative affect	21614.435	431				
	total perceived stress	14770.220	431				

a. R Squared = .000 (Adjusted R Squared = -.002)

b. R Squared = .008 (Adjusted R Squared = .006)

c. R Squared = .019 (Adjusted R Squared = .017)

		SEX		95% Confidence Interval	
Dependent Variable	SEX	Mean	Std. Error	Lower Bound	Upper Bound
Total Positive Affect	MALES	33.625	.534	32.575	34.675
	FEMALES	33.690	.460	32.785	34.594
Total Negative Affect	MALES	18.707	.521	17.683	19.730
	FEMALES	19.984	.448	19.103	20.865
Total perceived stress	MALES	25.788	.428	24.947	26.629
	FEMALES	27.419	.369	26.695	28.144

Interpretation of output from MANOVA

The key aspects of the output generated by MANOVA are presented below.

Descriptive statistics

Check that the information is correct. In particular, check that the N values correspond to what you know about your sample. These N values are your 'cell sizes' (see Assumption 1). Make sure that you have more subjects (cases) in each cell than the number of dependent variables. If you have over 30, then any violations of normality or equality of variance that may exist are not going to matter too much.

Box's Test

The output box labelled **Box's Test of Equality of Covariance Matrices** will tell you whether your data violates the assumption of homogeneity of variance-covariance matrices. If the **Sig.** value is *larger* than .001, then you have *not* violated the assumption. Tabachnick and Fidell (2001, p. 80) warn that Box's M can tend to be too strict when you have a large sample size. Fortunately, in our example the Box's M sig. value is .33; therefore, we have not violated this assumption.

Levene's Test

The next box to look at is **Levene's Test of Equality of Error Variances**. In the Sig. column look for any values that are *less* than .05. These would indicate that you have violated the assumption of equality of variance for that variable. In the current example, none of the variables recorded significant values; therefore, we can assume equal variances. If you do violate this assumption of equality of variances you will need to set a more conservative alpha level for determining significance for that variable in the univariate F-test. Tabachnick and Fidell (2001) suggest an alpha of .025 or .01, rather than the conventional .05 level.

Multivariate tests

This set of multivariate tests of significance will indicate whether there are statistically significant differences among the groups on a linear combination of the dependent variables. There are a number of statistics to choose from (Wilks' Lambda, Hotelling's Trace, Pillai's Trace). One of the most commonly reported statistics is **Wilks' Lambda**. Tabachnick and Fidell (2001) recommend Wilks' Lambda for general use; however, if your data have problems (small sample size, unequal N values, violation of assumptions), then Pillai's trace is more robust (see comparison of statistics in Tabachnick & Fidell, 2001, p. 348). In situations where you have only two groups, the F-tests for Wilks' Lambda, Hotelling's Trace and Pillai's Trace are identical.

Wilks' Lambda

You will find the value you are looking for in the *second* section of the **Multivariate Tests** table, in the row labelled with the name of your independent or grouping variable (in this case: SEX). Don't make the mistake of using the first set of figures, which refers to the intercept. Find the value of **Wilks' Lambda** and its associated significance level (**Sig.**). If the significance level is *less* than .05, then you can conclude that there is a difference among your groups. In the example shown above, we obtained a Wilks' Lambda value of .976, with a significance value of .014. This is less than .05; therefore, there is a statistically significant difference between males and females in terms of their overall wellbeing.

Between-subjects effects

If you obtain a significant result on this multivariate test of significance, this gives you permission to investigate further in relation to each of your dependent variables. Do males and females differ on all of the dependent measures, or just some? This information is provided in the **Tests of Between-Subjects Effects** output box. Because you are looking at a number of separate analyses here, it is suggested that you set a higher alpha level to reduce the chance of a Type 1 error (i.e. finding a significant result when there isn't really one). The most common way of doing this is to apply what is known as a Bonferroni adjustment. In its simplest form this involves dividing your original alpha level of .05 by the number of analyses that you intend to do (see Tabachnick & Fidell, 2001, p. 349, for more sophisticated versions of this formula). In this case we have three dependent variables to investigate; therefore, we would divide .05 by 3, giving a new alpha level of .017. We will consider our results significant only if the probability value (Sig.) is less than .017.

Significance

In the **Tests of Between-Subjects Effects** box, move down to the third set of values in a row labelled with your independent variable (in this case SEX). You will see each of your dependent variables listed, with their associated univariate F, df and Sig. values. You interpret these in the same way as you would a normal one-way analysis of variance. In the **Sig.** column look for any values that are less than .017 (our new adjusted alpha level). In our example, only one of the dependent variables (Total perceived stress) recorded a significance value less than our cut-off (with a Sig. value of .004). In this study, the only significant difference between males and females was on their perceived stress scores.

Effect size

The importance of the impact of sex on perceived stress can be evaluated using the effect size statistic provided by SPSS: **Partial Eta Squared**. Partial eta squared represents the proportion of the variance in the dependent variable (perceived stress scores) that can be explained by the independent variable (sex). The value in this case is .019, which, according to generally accepted criteria (Cohen, 1988), is considered quite a small effect (see the introduction to Part Five for a discussion of effect size). This represents only 1.9 per cent of the variance in perceived stress scores explained by sex.

Comparing group means

Although we know that males and females differed in terms of perceived stress, we do not know who had the higher scores. To find this out we refer to the output table provided in the section labelled **Estimated Marginal Means**. For Total perceived stress the mean score for males was 25.79 and for females 27.42. Although statistically significant, the actual difference in the two mean scores was very small, fewer than 2 scale points.

Follow-up analyses

In the example shown above there were only two levels to the independent variable (resulting in two groups: males, females). When you have independent variables with three or more levels it is necessary to conduct follow-up analyses to identify where the significant differences lie (is Group 1 different from Group 2, is Group 2 different from Group 3 etc.). These analyses are beyond the scope of this book. I suggest you consult a text on multivariate statistics (a number of recommended readings are given in the References sections).

Presenting the results from MANOVA

The results of this multivariate analysis of variance could be presented as follows:

A one-way between-groups multivariate analysis of variance was performed to investigate sex differences in psychological wellbeing. Three dependent variables were used: positive affect, negative affect and perceived stress. The independent variable was gender. Preliminary assumption testing was conducted to check for normality, linearity, univariate and multivariate outliers, homogeneity of variance-covariance matrices, and multicollinearity, with no serious violations noted. There was a statistically significant difference between males and females on the combined dependent variables: $F(3, 428)=3.57$, $p=.014$; Wilks' Lambda=.98; partial eta squared=.02. When the results for the dependent variables were considered separately, the only difference to reach statistical significance, using a Bonferroni adjusted alpha level of .017, was perceived stress: $F(1, 430)=8.34$, $p=.004$, partial eta squared=.02. An inspection of the mean scores indicated that females reported slightly higher levels of perceived stress ($M=27.42$, $SD=6.08$) than males ($M=25.79$, $SD=5.41$).

For other examples of the use of MANOVA and the presentation of results in journal articles, see Hair, Anderson, Tatham and Black (1998). This book is particularly useful if you are interested in seeing examples from business.

Additional exercise

Health

Data file: *sleep.sav*. See Appendix for details of the data file.

1. Conduct a one-way MANOVA to see if there are gender differences in each of the individual items that make up the Sleepiness and Associated Sensations Scale. The variables you will need as dependent variables are *fatigue, lethargy, tired, sleepy, energy*.

References

In a book such as this it is not appropriate for me to cover the complexity of the theory underlying MANOVA or all its associated assumptions and techniques. I would very strongly recommend that you read up on MANOVA and make sure that you understand it. There are a number of good books that you might like to use for this purpose.

Strange looking numbers

In your output you might come across strange looking numbers which take the form: 1.24E–02. These numbers have been written in scientific notation. They can easily be converted to a 'normal' number for presentation in your results. The number after the E tells you how many places you need to shift the decimal point. If the number after the E is negative, shift the decimal point to the left; if it is positive, shift it to the right. Therefore, 1.24E–02 becomes 0.124. The number 1.24E02 would be written as 124.

You can prevent these small numbers from being displayed in this strange way. Go to **Edit**, **Options** and put a tick in the box labelled: **No scientific notation for small numbers in tables**. Rerun the analysis and normal numbers should appear in all your output.

Cohen, J. (1988). *Statistical power analysis for the behavioral sciences*. Hillsdale, NJ: Erlbaum.

Hair, J. F., Anderson, R. E., Tatham, R. L., & Black, W. C. (1998). *Multivariate data analysis* (5th edn). New York: Macmillan.

Stevens, J. (1996). *Applied multivariate statistics for the social sciences* (3rd edn). Mahway, NJ: Lawrence Erlbaum.

Tabachnick, B. G., & Fidell, L. S. (2001). *Using multivariate statistics* (4th edn). New York: HarperCollins.

21 Analysis of covariance

Analysis of covariance is an extension of analysis of variance (discussed in Chapter 20) that allows you to explore differences between groups while statistically controlling for an additional (continuous) variable. This additional variable (called a covariate) is a variable that you suspect may be influencing scores on the dependent variable. SPSS uses regression procedures to remove the variation in the dependent variable that is due to the covariate/s, and then performs the normal analysis of variance techniques on the corrected or adjusted scores. By removing the influence of these additional variables ANCOVA can increase the power or sensitivity of the F-test. That is, it may increase the likelihood that you will be able to detect differences between your groups.

Analysis of covariance can be used as part of one-way, two-way and multivariate ANOVA techniques. In this chapter SPSS procedures are discussed for analysis of covariance associated with the following designs:

- one-way between-groups ANOVA (1 independent variable, 1 dependent variable); and
- two-way between-groups ANOVA (2 independent variables, 1 dependent variable).

In the following sections you will be briefly introduced to some background on the technique and assumptions that need to be tested; then a number of worked examples are provided. Before running the analyses I suggest you read the introductory sections in this chapter, and revise the appropriate ANOVA chapters (Chapter 17 and 18). Multivariate ANCOVA is not illustrated in this chapter. If you are interested in this technique, I suggest you read Tabachnick and Fidell (2001).

Uses of ANCOVA

ANCOVA can be used when you have a two-group pre-test/post-test design (e.g. comparing the impact of two different interventions, taking before and after measures for each group). The scores on the pre-test are treated as a covariate to 'control' for pre-existing differences between the groups. This makes ANCOVA very useful in situations when you have quite small sample sizes, and only small or medium effect sizes (see discussion on effect sizes in the introduction to Part

Five). Under these circumstances (which are very common in social science research), Stevens (1996) recommends the use of two or three carefully chosen covariates to reduce the error variance and increase your chances of detecting a significant difference between your groups.

ANCOVA is also handy when you have been unable to randomly assign your subjects to the different groups, but instead have had to use existing groups (e.g. classes of students). As these groups may differ on a number of different attributes (not just the one you are interested in), ANCOVA can be used in an attempt to reduce some of these differences. The use of well-chosen covariates can help reduce the confounding influence of group differences. This is certainly not an ideal situation, as it is not possible to control for all possible differences; however, it does help reduce this systematic bias. The use of ANCOVA with intact or existing groups is somewhat of a contentious one among writers in the field. It would be a good idea to read more widely if you find yourself in this situation. Some of these issues are summarised in Stevens (1996, pp. 324–327).

Choosing appropriate covariates

ANCOVA can be used to control for one or more covariates at the same time. These covariates need to be chosen carefully, however (see Stevens, 1996, p. 320; and Tabachnick & Fidell, 2001, p. 279). In identifying possible covariates you should ensure you have a good understanding of the theory and previous research that has been conducted in your topic area. The variables that you choose as your covariates should be continuous variables, measured reliably (see Chapter 9), and should correlate significantly with the dependent variable. Ideally, you should choose a small set of covariates that are only moderately correlated with one another, so that each contributes uniquely to the variance explained. The covariate must be measured before the treatment or experimental manipulation is performed. This is to prevent scores on the covariate from also being influenced by the treatment.

Alternatives to ANCOVA

There are a number of assumptions or limitations associated with ANCOVA. Sometimes you will find that your research design or your data are not suitable. Both Tabachnick and Fidell (2001, p. 303–304) and Stevens (1996, p. 327) suggest a number of alternative approaches to ANCOVA. It would be a good idea to explore these alternatives and to evaluate the best course of action, given your particular circumstances.

Assumptions of ANCOVA

There are a number of issues and assumptions associated with ANCOVA. These are over and above the usual ANOVA assumptions discussed in the introduction to Part Five. In this chapter only the key assumptions for ANCOVA will be discussed. Tabachnick and Fidell (2001, pp. 280–283) have a good, detailed coverage of this topic, including the issues of unequal sample sizes, outliers, multicollinearity, normality, homogeneity of variance, linearity, homogeneity of regression and reliability of covariates (what an impressive and awe-inspiring list of statistical jargon!).

Influence of treatment on covariate measurement

In designing your study you should ensure that the covariate is measured *prior to* the treatment or experimental manipulation. This is to avoid scores on the covariate also being influenced by the treatment. If the covariate is affected by the treatment condition, then this change will be correlated with the change that occurs in your dependent variable. When ANCOVA removes (controls for) the covariate, it will also remove some of the treatment effect, thereby reducing the likelihood of obtaining a significant result.

Reliability of covariates

ANCOVA assumes that covariates are measured without error, which is a rather unrealistic assumption in much social science research. Some variables that you may wish to control, such as age, can be measured reasonably reliably; others which rely on a scale may not meet this assumption. There are a number of things you can do to improve the reliability of your measurement tools:

- Look for good, well-validated scales and questionnaires. Make sure they measure what you think they measure (don't just rely on the title—check the manual and inspect the items).
- Check that the measure you intend to use is suitable for use with your sample (some scales may be very reliable for adults, but may not be suitable for children or adolescents).
- Check the internal consistency (a form of reliability) of your scale by calculating Cronbach alpha (see Chapter 9). Values should be above .7 or .8 to be considered reliable. Be careful with very short scales (e.g. under ten items), however, because Cronbach alpha is quite sensitive to the number of items in the scale. Short scales often have low Cronbach alpha values.

In this case you may need to check the inter-correlations among the scale items.

- If you have had to write the questions yourself, make sure they are clear, appropriate and unambiguous. Make sure the questions and response scales are appropriate for all of your groups (e.g. consider sex differences, language difficulties, cultural backgrounds). Always pilot-test your questions before conducting the full study.

- Consider the circumstances under which you are measuring your covariate. Are people likely to answer honestly, or are they likely to distort their answers to avoid embarrassment, etc.?

- If you are using any form of equipment or measuring instrumentation, make sure that it is functioning properly and calibrated appropriately. Make sure the person operating the equipment is competent and trained in its use.

- If your study involves using other people to observe or rate behaviour, make sure they are trained and that each observer uses the same criteria. Preliminary pilot-testing to check inter-rater consistency would be useful here.

Correlations among covariates

There should not be strong correlations among the variables you choose for your covariates. Ideally, you want a group of covariates that correlate substantially with the dependent variable but not with one another. To choose appropriate covariates you will need to use the existing theory and research to guide you. Run some preliminary correlation analyses (see Chapter 11) to explore the strength of the relationship among your proposed covariates. If you find that the covariates you intend to use correlate strongly (e.g. r=.80), you should consider removing one or more of them (see Stevens, 1996, p. 320). Each of the covariates you choose should pull their own weight—overlapping covariates do not contribute to a reduction in error variance.

Linear relationship between dependent variable and covariate

ANCOVA assumes that the relationship between the dependent variable and each of your covariates is linear (straight-line). If you are using more than one covariate, it also assumes a linear relationship between each of the pairs of your covariates. Violations of this assumption are likely to reduce the power (sensitivity) of your test. Remember, one of the reasons for including covariates was to increase the power of your analysis of variance test.

Scatterplots can be used to test for linearity, but these need to be checked separately for each of your groups (i.e. the different levels of your independent variable). If you discover any curvilinear relationships, these may be corrected by transforming your variable (see Chapter 8), or alternatively you may wish to drop the offending covariate from the analysis. Disposing of covariates that misbehave is often easier, given the difficulty in interpreting transformed variables.

Homogeneity of regression slopes

This impressive-sounding assumption requires that the relationship between the covariate and dependent variable for each of your groups is the same. This is indicated by similar slopes on the regression line for each group. Unequal slopes would indicate that there is an interaction between the covariate and the treatment. If there is an interaction, then the results of ANCOVA are misleading, and therefore it should not be conducted (see Stevens, 1996, pp. 323, 331; Tabachnick & Fidell, 2001, p. 282). The procedure for checking this assumption is provided in the examples presented later in this chapter.

One-way ANCOVA

In this section you will be taken step by step through the process of performing a one-way analysis of covariance. One-way ANCOVA involves one independent, categorical variable (with two or more levels or conditions), one dependent continuous variable, and one or more continuous covariates. This technique is often used when evaluating the impact of an intervention or experimental manipulation, while controlling for pre-test scores.

Details of example

To illustrate the use of one-way ANCOVA, I will be using the experim.sav data file included on the website that accompanies this book (see p. xi). These data refer to a fictitious study that involves testing the impact of two different types of interventions in helping students cope with their anxiety concerning a forthcoming statistics course (see the Appendix for full details of the study). Students were divided into two equal groups and asked to complete a number of scales (including one that measures fear of statistics). One group was given a number of sessions designed to improve their mathematical skills, the second group participated in a program designed to build their confidence. After the program they were again asked to complete the same scales they completed before the program. The manufactured data file is included on the disk provided with this book. If you wish to follow the procedures detailed below you will need to start SPSS, and open the data file labelled experim.sav. This file can be opened only in SPSS.

In this example I will explore the impact of the Maths Skills class (Group 1) and the Confidence Building class (Group 2) on participants' scores on the Fear of Statistics test, while controlling for the scores on this test administered before the program. Details of the variables names and labels from the data file are as follows:

File name	Variable name	Variable label	Coding instructions
experim.sav	Group	Type of class	1=Maths Skills 2=Confidence Building
	Fost1	Fear of Statistics scores at time1	Total scores on the Fear of Statistics test administered prior to the program. Scores range from 20 to 60. High scores indicate greater fear of statistics.
	Fost2	Fear of Statistics scores at time2	Total scores on the Fear of Statistics test administered after the program was complete. Scores range from 20 to 60. High scores indicate greater fear of statistics.
	Fost3	Fear of Statistics scores at time3	Total scores on the Fear of Statistics test administered 3 months after the program was complete. Scores range from 20 to 60. High scores indicate greater fear of statistics.

Summary for one-way ANCOVA

Example of research question: Is there a significant difference in the Fear of Statistics test scores for the maths skills group (group1) and the confidence building group (group2), while controlling for their pre-test scores on this test?

What you need: At least three variables are involved:
- one categorical independent variable with two or more levels (group1/group2);
- one continuous dependent variable (scores on the Fear of Statistics test at Time 2); and
- one or more continuous covariates (scores on the Fear of Statistics test at Time 1).

What it does: ANCOVA will tell us if the mean Fear of Statistics test scores at Time 2 for the two groups are significantly different after the initial pre-test scores are controlled for.

Assumptions: All normal one-way ANOVA assumptions apply (see the introduction to Part Five). These should be checked first.

Additional ANCOVA assumptions (see description of these presented earlier in this chapter):

1. the covariate is measured prior to the intervention or experimental manipulation;

2. the covariate is measured without error (or as reliably as possible);

3. the covariates are not strongly correlated with one another;

4. there is a linear relationship between the dependent variable and the covariate for all groups (linearity); and

5. the relationship between the covariate and dependent variable is the same for each of the groups (homogeneity of regression slopes).

Non-parametric alternative: none

Testing assumptions

Before you can begin testing the specific assumptions associated with ANCOVA, you will need to check the assumptions for a normal one-way analysis of variance (normality, homogeneity of variance). You should review the introduction to Part Five, and Chapter 17, before going any further here. To save space I will not repeat that material here, I will just discuss the additional ANCOVA assumptions listed in the summary above.

Assumption 1: Measurement of the covariate

This assumption specifies that the covariate should be measured before the treatment or experimental manipulation begins. This is not tested statistically, but instead forms part of your research design and procedures. This is why it is important to plan your study with a good understanding of the statistical techniques that you intend to use.

Assumption 2: Reliability of the covariate

This assumption concerning the reliability of the covariate is also part of your research design, and involves choosing the most reliable measuring tools available. If you have used a psychometric scale or measure, you can check the internal consistency reliability by calculating Cronbach alpha using the SPSS **Reliability** procedures (see Chapter 9). Given that I have 'manufactured' the data to illustrate this technique, I cannot test the reliability of the Fear of Statistics test. If they were real data I would check that the Cronbach alpha value was at least .70 (preferably .80).

Assumption 3: Correlations among the covariates

If you are using more than one covariate, you should check to see that they are not too strongly correlated with one another (r=.8 and above). To do this you will need to use the SPSS **Correlation** procedure (this is described in detail in Chapter 11). As I have only one covariate, I do not need to do this here.

Assumption 4: Linearity

There are a number of different ways that you can check the assumption of a linear relationship between the dependent variable and the covariates for all your groups. One way would be to (a) split your sample according to the various groups in your design using the SPSS **Split File** option, and then (b) generate scatterplots between the dependent variable and each of the covariates. In the procedure section below I will show you an alternative means. This involves generating just one scatterplot, but asking SPSS to show your groups separately.

Procedure for checking linearity for each group

Step 1:

1. From the menu at the top of the screen click on: **Graphs**, then click on **Scatter**.

2. Click on **Simple**. Click on the **Define** button.

3. In the **Y axis** box put your dependent variable (in this case: fear of stats time2)

4. In the **X axis** box put your covariate (e.g. fear of stats time1)

5. In the **Set markers by** box put your independent variable (e.g. group). Click on **OK**.

Step 2:

1. Once you have your scatterplot displayed in the **Viewer** window, double click on it to open the **Chart Editor** window.

2. In the legend, click on the first group listed: Maths Skills. With this highlighted move your cursor to the menu bar and click on the icon that looks like a scatterplot with a line drawn through it (Add fit line). A regression line for the Maths Skills group should appear on your scatterplot. Repeat the process to get a separate regression line for the Confidence Building group.

3. To show the lines more clearly when you print this out you might like to change one of the lines to a dotted line. Click on one of the lines on the scatterplot, right click on your mouse to open the **Properties** window. Click on the **Lines** tab. In the section labelled **Style** click on the drop down arrow and choose a dotted line. To use a thicker line, click in the section **Weight** and choose a larger value (e.g. 3). Click on **Apply** and then **Close**.

4. To close the graph and return to your normal **Viewer** window, click on **File** and then **Close**.

The output from this procedure is shown below.

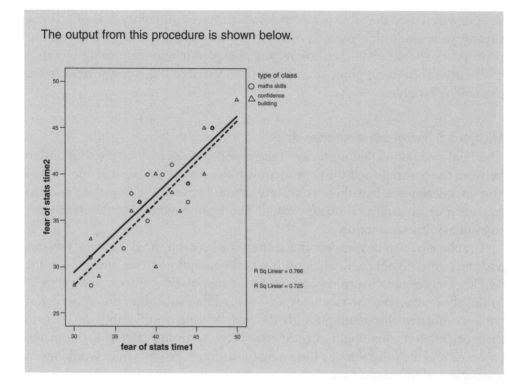

The output generated using the procedures detailed above provides you with a number of useful pieces of information.

First, you can look at the general distribution of scores for each of your groups. Does there appear to be a linear (straight-line) relationship for each group? What you don't want to see is an indication of a curvilinear relationship. In the above example the relationship is clearly linear, so we have not violated the assumption of a linear relationship. If you find a curvilinear relationship you may want to reconsider the use of this covariate, or alternatively you could try transforming the variable (see Chapter 8) and repeating the scatterplot to see whether there is an improvement.

The R squared values, which are given in the legend for each group, gives you an indication of the strength of the relationship between your dependent variable (Fear of Stats 2) and your covariate (Fear of Stats 1). In this case these two variables are strongly correlated. For the confidence building group, 76.6 per cent of the variance in scores at Time 2 are explained by scores at Time 1. (To get this value I just converted my R squared value to a percentage by shifting the decimal point two places.) For the maths skills group 72.5 per cent of the variance in the two sets of scores are shared. In many studies, particularly when you are using different scales as covariates, you will not find such high correlations. The high correlation here is due in part to the pre-intervention, post-intervention design, which used the same test administered on two different occasions. If you find in your study that there is only a very weak relationship between your dependent variables and the covariate, you may need to consider using alternative covariates. There is no point controlling for a variable if it is not related to the dependent variable.

Assumption 5: Homogeneity of regression slopes

This final assumption (homogeneity of regression slopes) concerns the relationship between the covariate and the dependent variable for each of your groups. What you are checking is that there is no interaction between the covariate and the treatment or experimental manipulation. There are a number of different ways to evaluate this assumption.

Graphically, you can inspect the scatterplot between the dependent variable and the covariate obtained when testing for Assumption 4 (see above). Are the two lines (corresponding to the two groups in this study) similar in their slopes? In the above example the two lines are very similar, so it does not appear that we have violated this assumption. If the lines had been noticeably different in their orientation, this might suggest that there is an interaction between the covariate and the treatment (as shown by the different groups). This would mean a violation of this assumption.

This assumption can also be assessed statistically, rather than graphically. This involves checking to see whether there is a statistically significant interaction between the treatment and the covariate. If the interaction is significant at an alpha level of .05, then we have violated the assumption. The procedure to check this is described next.

Procedure to check for homogeneity of regression slopes

1. From the menu at the top of the screen click on: **Analyze**, then click on **General Linear Model**, then on **Univariate**.

2. In the **Dependent Variables** box put your dependent variable (e.g. fear of stats time2, abbreviated to fost2).

3. In the **Fixed Factor** box put your independent or grouping variable (e.g. type of class, abbreviated to group).

4. In the **Covariate** box put your covariate (e.g. fear of stats time1, abbreviated to fost1).

5. Click on the **Model** button. Click on **Custom**.

6. Check that the **Interaction** option is showing in the **Build Terms** box.

7. Click on your independent variable (Group) and then the arrow button to move it into the **Model** box.

8. Click on your covariate (fost1) and then the arrow button to move it into the **Model** box.

9. Go back and click on your independent variable (Group) again on the left-hand side (in the Factors and Covariates section). While this is highlighted, hold down the Ctrl button on your keyboard, and then click on your covariate variable (fost1). Click on the arrow button to move this into the right-hand side box labelled **Model**.

10. In the **Model** box you should now have listed:

- your independent variable (Group);
- your covariate (Fost1); and
- an extra line of the form: covariate * independent variable (group*fost1).

11. The final term is the interaction that we are checking for.

12. Click on **Continue** and then **OK**.

The output generated by this procedure is shown below.

Tests of Between-Subjects Effects

Dependent Variable : fear of stats time2

Source	Type III Sum of Squares	df	Mean Square	F	Sig.
Corrected Model	577.689 [a]	3	192.563	26.102	.000
Intercept	3.877	1	3.877	.525	.475
GROUP	.846	1	.846	.115	.738
FOST1	538.706	1	538.706	73.022	.000
GROUP* FOST1	.409	1	.409	.055	.816
Error	191.811	26	7.377		
Total	42957.000	30			
Corrected Total	769.500	29			

[a.] R Squared = .751 (Adjusted R Squared = .722)

In the output obtained from this procedure, the only value that you are interested in is the significance level of the interaction term (shown above as Group*Fost1). You can ignore the rest of the output. If the Sig. level for the interaction is less than or equal to .05, then your interaction is statistically significant, indicating that you have violated the assumption. In this situation we do *not* want a significant result. We want a Sig. value of *greater* than .05. In the above example the Sig or probability value is .816, safely above the cut-off. We have not violated the assumption of homogeneity of regression slopes. This supports the earlier conclusion gained from an inspection of the scatterplots for each group.

Now that we have finished checking the assumptions, we can proceed with the ANCOVA analysis to explore the differences between our treatment groups.

Procedure for one-way ANCOVA

1. From the menu at the top of the screen click on: **Analyze**, then click on **General Linear Model**, then on **Univariate**.

2. In the **Dependent Variables** box put your dependent variable (e.g. fear of stats time2, abbreviated to fost2).

3. In the **Fixed Factor** box put your independent or grouping variable (e.g. type of class, abbreviated to group).

4. In the **Covariate** box put your covariate (e.g. fear of stats time1, abbreviated to fost1).

5. Click on the **Model** button. Click on **Full Factorial** in the **Specifify Model** section. Click on **Continue**.

6. Click on the **Options** button.

7. In the top section labelled **Estimated Marginal Means** click on your independent variable (group).

8. Click on the arrow to move it into the box labelled **Display Means for**. This will provide you with the mean score on your dependent variable for each group, adjusted for the influence of the covariate.

9. In the bottom section of the **Options** dialogue box choose **Descriptive statistics**, **Estimates of effect size** and **Homogeneity tests**.

10. Click on **Continue** and then **OK**.

The output generated from this procedure is shown below.

Descriptive Statistics

Dependent Variable: fear of stats time2

type of class	Mean	Std. Deviation	N
maths skills	37.67	4.51	15
confidence building	37.33	5.88	15
Total	37.50	5.15	30

Levene's Test of Equality of Error Variances[a]

Dependent Variable: fear of stats time2

F	df1	df2	Sig.
.141	1	28	.710

Tests the null hypothesis that the error variance of the
dependent variable is equal across groups.

a. Design: Intercept +FOST1+GROUP

Tests of Between-Subjects Effects

Dependent Variable: fear of stats time2

Source	Type III Sum of Squares	df	Mean Square	F	Sig.	Partial Eta Squared
Corrected Model	577.279[a]	2	288.640	40.543	.000	.750
Intercept	3.510	1	3.510	.493	.489	.018
fost1	576.446	1	576.446	80.970	.000	.750
group	5.434	1	5.434	.763	.390	.027
Error	192.221	27	7.119			
Total	42957.000	30				
Corrected Total	769.500	29				

a. R Squared = .750 (Adjusted R Squared = .732)

type of class

Dependent Variable: fear of stats time2

type of class	Mean	Std. Error	95% Confidence Interval Lower Bound	95% Confidence Interval Upper Bound
maths skills	37.926[a]	.690	36.512	39.341
confidence building	37.074[a]	.690	35.659	38.488

a. Evaluated at covariates appeared in the model: fear of stats time1 = 40.17.

Interpretation of output from one-way ANCOVA

There are a number of steps in interpreting the output from ANCOVA.

- In the box labelled **Descriptive Statistics,** check that the details are correct (e.g. number in each group, mean scores).
- The details in the table labelled Levene's Test of Equality of Error Variances allow you to check that you have not violated the assumption of equality of variance. You want the Sig. value to be *greater* than .05. If this value is smaller than .05 (and therefore significant), this means that your variances are not equal, and that you have violated the assumption. In this case we have not violated the assumption because our Sig. value is .71, which is much larger than our cut-off of .05.

• The main ANCOVA results are presented in the next table, labelled **Test of Between-Subjects Effects**. We want to know whether our groups are significantly different in terms of their scores on the dependent variable (e.g. Fear of Statistics Time 2). Find the line corresponding to your independent variable (in this case Group) and read across to the column labelled **Sig.** If the value in this column is *less* than .05 (or an alternative alpha level you have set), then your groups differ significantly. In this case our value is .39, which is greater than .05; therefore, our result is *not* significant. There is not a significant difference in the fear of statistics scores for subjects in the maths skills group and the confidence building group, after controlling for scores on the Fear of Statistics test administered prior to the intervention.

• You should also consider the effect size, as indicated by the corresponding partial eta squared value (see the introduction to Part Five for a description of what an effect size is). The value in this case is only .027 (a small effect size according to Cohen's 1988 guidelines). This value also indicates how much of the variance in the dependent variable is explained by the independent variable. Convert the partial eta squared value to a percentage by multiplying by 100 (shift the decimal point two places to the right). In this example we are able to explain only 2.7 per cent of the variance.

• The other piece of information that we can gain from this table concerns the influence of our covariate. Find the line in the table that corresponds to the covariate (e.g. FOST1: Fear of Statistics at time1). Read across to the Sig. level. This indicates whether there is a significant relationship between the covariate and the dependent variable, while controlling for the independent variable (group). In the line corresponding to Fost1 (our covariate) you will see that the Sig. value is .000 (which actually means less than .0005). This is less than .05, so our covariate is significant. In fact, it explained 75 per cent of the variance in the dependent variable (partial eta squared of .75 multiplied by 100).

• The final table in the ANCOVA output (**Estimated marginal means**) provides us with the adjusted means on the dependent variable for each of our groups. 'Adjusted' refers to the fact that the effect of the covariate has been statistically removed.

Presenting the results from one-way ANCOVA

The results of this one-way analysis of covariance could be presented as follows:

A one-way between-groups analysis of covariance was conducted to compare the effectiveness of two different interventions designed to reduce participants' fear of statistics. The independent variable was the type of intervention (maths skills, confidence building), and the dependent variable consisted of scores on the Fear of Statistics test administered after the intervention was completed. Participants' scores on the pre-intervention administration of the Fear of Statistics test were used

Strange looking numbers

In your output you might come across strange looking numbers which take the form: 1.24E–02. These numbers have been written in scientific notation. They can easily be converted to a 'normal' number for presentation in your results. The number after the E tells you how many places you need to shift the decimal point. If the number after the E is negative, shift the decimal point to the left; if it is positive, shift it to the right. Therefore, 1.24E–02 becomes 0.124. The number 1.24E02 would be written as 124.

You can prevent these small numbers from being displayed in this strange way. Go to **Edit, Options** and put a tick in the box labelled: **No scientific notation for small numbers in tables**. Rerun the analysis and normal numbers should appear in all your output.

as the covariate in this analysis. Preliminary checks were conducted to ensure that there was no violation of the assumptions of normality, linearity, homogeneity of variances, homogeneity of regression slopes, and reliable measurement of the covariate. After adjusting for pre-intervention scores, there was no significant difference between the two intervention groups on post-intervention scores on the Fear of Statistics test [$F(1,27)=.76$, $p=.39$, partial eta squared=.03]. There was a strong relationship between the pre-intervention and post-intervention scores on the Fear of Statistics test, as indicated by a partial eta squared value of .75.

In presenting the results of this analysis you would also provide a table of means for each of the groups. If the same scale is used to measure the dependent variable (Time 2) and the covariate (Time 1), then you will include the means at Time 1 and Time 2. You can get these by running **Descriptives** (see Chapter 6).

If a different scale is used to measure the covariate, you will provide the 'unadjusted' mean (and standard deviation), and the 'adjusted' mean (and standard error) for the two groups. The unadjusted mean is available from the Descriptive Statistics table. The adjusted mean (controlling for the covariate) is provided in the Estimated marginal means table. It is also a good idea to include the number of cases in each of your groups.

Two-way ANCOVA

In this section you will be taken step by step through the process of performing a two-way analysis of covariance. Two-way ANCOVA involves two independent, categorical variables (with two or more levels or conditions), one dependent continuous variable, and one or more continuous covariates. It is important that you have a good understanding of the standard two-way analysis of variance procedure and assumptions before proceeding further here. Refer to Chapter 18 in this book.

The example that I will use here is an extension of that presented in the one-way ANCOVA section above. In that analysis I was interested in determining which intervention (maths skills or confidence building) was more effective in reducing students' fear of statistics. I found no significant difference between the groups.

Suppose that in reading further in the literature on the topic I found some research that suggested there might be a difference in how males and females respond to different interventions. Sometimes in the literature you will see this additional variable (e.g. sex) described as a *moderator*. That is, it moderates or influences the effect of the other independent variable. Often these moderator variables are individual difference variables, characteristics of individuals that influence the way in which they respond to an experimental manipulation or treatment condition. It is important if you obtain a non-significant result for your one-way ANCOVA that you consider the possibility of moderator variables. Some of the most interesting research occurs when a researcher stumbles across (or systematically investigates) moderator variables that help to explain why some researchers obtain statistically significant results while others do not. In your own research, always consider factors such as gender and age, as these can play an important part in influencing the results. Studies conducted on young university students often don't generalise to broader (and older) community samples. Research on males sometimes yields quite different results when repeated using female samples. The message here is to consider the possibility of moderator variables in your research design and, where appropriate, include them in your study.

Details of example

In the example presented below I will use the same data that were used in the previous section, but I will add an additional independent variable (gender). This will allow me to broaden my analysis to see whether gender is acting as a moderator variable in influencing the effectiveness of the two programs. I am interested in the possibility of a significant interaction effect between sex and intervention group. Males might benefit most from the maths skills intervention, while females might benefit more from the confidence building program.

Warning: The sample size used to illustrate this example is very small, particularly when you break the sample down by the categories of the independent variables (gender and group). If using this technique with your own research, you should really try to have a much larger data file overall, with a good number in each of the categories of your independent variables.

If you wish to follow along with the steps detailed below you will need to start SPSS and open the data file labelled experim.sav provided on the website accompanying this book (see p. xi). This file can be opened only in SPSS. (Remember, this is just a manufactured data file, so don't get too excited about the results—they are designed merely to illustrate the use of the procedure.) See the table below for the variables used.

File name	Variable name	Variable label	Coding instructions
Experim.sav	Group	Type of class	1=Maths Skills 2=Confidence Building
	Sex	Sex	1=male 2=female
	Fost1	Fear of Statistics scores at time1	Total scores on the Fear of Statistics test administered prior to the program. Scores range from 20 to 60. High scores indicate greater fear of statistics.
	Fost2	Fear of Statistics scores at time2	Total scores on the Fear of Statistics test administered after the program was complete. Scores range from 20 to 60. High scores indicate greater fear of statistics.

Summary for two-way ANCOVA

Example of research question: Does gender influence the effectiveness of two programs designed to reduce participants' fear of statistics?

Is there a difference in post-intervention Fear of Statistics test scores between males and females in their response to a maths skills program and a confidence building program?

What you need: At least four variables are involved:
- two categorical independent variables with two or more levels (sex: M/F; group: Maths skills/ Confidence building);
- one continuous dependent variable (Fear of Statistics test scores at Time 2); and
- one or more continuous covariates (Fear of Statistics test scores at Time 1).

What it does: ANCOVA will control for scores on your covariate/s and then perform a normal two-way ANOVA. This will tell you if there is:
- a significant main effect for your first independent variable (group);
- a main effect for your second independent variable (sex); and
- a significant interaction between the two.

Assumptions: All normal two-way ANOVA assumptions apply (e.g. normality, homogeneity of variance). These should be checked first (see the introduction to Part Five).

Additional ANCOVA assumptions: See discussion of these assumptions, and the procedures to test them, in the one-way ANCOVA section presented earlier.

Non-parametric alternative: none

Procedure for two-way ANCOVA

1. From the menu at the top of the screen click on: **Analyze**, then click on **General Linear Model**, then on **Univariate**.

2. Click on your dependent variable (e.g. fear of stats time2, abbreviated to fost2) and move it into the **Dependent Variables** box.

3. Click on your two independent or grouping variables (e.g. sex; group). Move these into the **Fixed Factor** box.

4. In the **Covariate** box put your covariate/s (e.g. fear of stats time1, abbreviated to fost1).

5. Click on the **Model** button. Click on **Full Factorial** in the **Specify Model** section. Click on **Continue**.

6. Click on the **Options** button.

7. In the top section labelled **Estimated Marginal Means** click on your first independent variable (e.g. sex). Click on the arrow to move it into the box labelled **Display means for**.

8. Repeat for the second independent variable (e.g. group).

9. Click on the extra interaction term included (e.g. group*sex). Move this into the box. This will provide you with the mean scores on your dependent variable split for each group, adjusted for the influence of the covariate.

10. In the bottom section of the **Options** screen click on **Descriptive statistics**, **Estimates of effect size** and **Homogeneity tests**. Click on **Continue**.

11. You might also like to see a graph of the relationship among your variables (optional). To do this Click on the **Plots** button.

12. Highlight your first independent variable (e.g. group) and move this into the box labelled **Horizontal**. This variable will appear across the bottom of your graph.

13. Click on your second independent variable (e.g. sex) and move this into the box labelled **Separate Lines**. This variable will be represented by different coloured lines for each group.

14. Click on **Add**. You should see listed in the bottom Plots box a formula that takes the form:

independent variable 1 (e.g. group) * independent variable 2 (e.g. sex).

15. Click on **Continue** and then **OK**.

The output generated from this procedure is shown below.

Descriptive Statistics

Dependent Variable : fear of stats time2

type of class	SEX	Mean	Std. Deviation	N
maths skills	male	37.25	5.50	8
	female	38.14	3.44	7
	Total	37.67	4.51	15
confidence building	male	40.57	5.56	7
	female	34.50	4.78	8
	Total	37.33	5.88	15
Total	male	38.80	5.60	15
	female	36.20	4.48	15
	Total	37.50	5.15	30

Levene's Test of Equality of Error Variances[a]

Dependent Variable: fear of stats time2

F	df1	df2	Sig.
2.204	3	26	.112

Tests the null hypothesis that the error variance of the dependent variable is equal across groups.

a. Design: Intercept+FOST1+GROUP+SEX+GROUP * SEX

Tests of Between-Subjects Effects

Dependent Variable: fear of stats time2

Source	Type III Sum of Squares	df	Mean Square	F	Sig.	Partial Eta Squared
Corrected Model	686.728[a]	4	171.682	51.854	.000	.892
Intercept	4.137	1	4.137	1.250	.274	.048
fost1	545.299	1	545.299	164.698	.000	.868
group	4.739	1	4.739	1.431	.243	.054
sex	4.202	1	4.202	1.269	.271	.048
group * sex	104.966	1	104.966	31.703	.000	.559
Error	82.772	25	3.311			
Total	42957.000	30				
Corrected Total	769.500	29				

a. R Squared = .892 (Adjusted R Squared = .875)

Estimated marginal means

1. type of class

Dependent Variable: fear of stats time2

type of class	Mean	Std. Error	95% Confidence Interval	
			Lower Bound	Upper Bound
maths skills	38.024[a]	.472	37.053	38.996
confidence building	37.226[a]	.471	36.255	38.197

a. Evaluated at covariates appeared in the model: fear of stats time1 = 40.17.

2. SEX

Dependent Variable: fear of stats time2

SEX	Mean	Std. Error	95% Confidence Interval	
			Lower Bound	Upper Bound
male	38.009[a]	.476	37.028	38.989
female	37.242[a]	.476	36.261	38.222

a. Evaluated at covariates appeared in the model: fear of stats time1 = 40.17.

3. type of class* SEX

Dependent Variable: fear of stats time2

type of class	SEX	Mean	Std. Error	95% Confidence Interval	
				Lower Bound	Upper Bound
maths skills	male	36.532[a]	.646	35.202	37.862
	female	39.517[a]	.696	38.083	40.950
confidence building	male	39.485[a]	.693	38.058	40.912
	female	34.966[a]	.644	33.639	36.294

a. Evaluated at covariates appeared in the model: fear of stats time1 = 40.17.

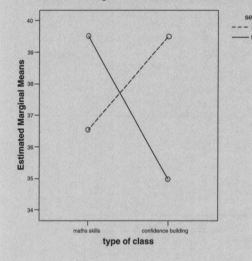

Estimated Marginal Means of fear of stats time2

Interpretation of output from two-way ANCOVA

There are a number of steps in interpreting the output from a two-way ANCOVA.

- In the box labelled **Descriptive Statistics,** check that the details are correct (e.g. number in each group, mean scores).
- The details in the table labelled **Levene's Test of Equality of Error Variances** allow you to check that you have not violated the assumption of equality of variance. You want the **Sig.** value to be *greater* than .05. If this value is smaller than .05 (and therefore significant), this means that your variances are not equal, and that you have violated the assumption. In this case we have not violated the assumption because our Sig. value is .11, which is much larger than our cut-off of .05.
- The main ANCOVA results are presented in the next table, labelled **Test of Between-Subjects Effects.** We want to know whether there is a significant main effect for any of our independent variables (groups or sex) and whether

the interaction between these two variables is significant. Of most interest is the interaction, so we will check this first. If the interaction is significant, then the two main effects are not important, because the effect of one independent variable is dependent on the level of the other independent variable. Find the line corresponding to the interaction effect (group* sex in this case). Read across to the column labelled **Sig.** If the value in this column is *less* than .05 (or an alternative alpha level you have set), then the interaction is significant. In this case our Sig. (or probability) value is .000 (read this as less than .0005), which is less than .05; therefore, our result is significant. This significant interaction effect suggests that males and females respond differently to the two programs. The main effects for Group and for Sex are not statistically significant (Group: p=.24; Sex: p=.27). We cannot say that one intervention is better than the other, because we must consider whether we are referring to males or females. We cannot say that males benefit more than females, because we must consider which intervention was involved.

- You should also consider the effect size, as indicated by the corresponding partial eta squared value (see the introduction to Part Five for a description of what an effect size is). The value in this case is .56 (a large effect size according to Cohen's 1988 guidelines). This value also indicates how much of the variance in the dependent variable is explained by the independent variable. Convert the partial eta squared value to a percentage by multiplying by 100 (shift the decimal point two places to the right). In this example we are able to explain 56 per cent of the variance.

- The other piece of information that we can gain from this table concerns the influence of our covariate. Find the line in the table that corresponds to the covariate (e.g. FOST1: Fear of Statistics at time1). Read across to the Sig. level. This indicates whether there is a significant relationship between the covariate and the dependent variable, while controlling for the independent variable (group). In the line corresponding to Fost1 (our covariate) you will see that the Sig. value is .000 (which actually means less than .0005). This is less than .05; therefore, our covariate was significant. In fact, it explained 87 per cent of the variance in the dependent variable (partial eta squared of .87 multiplied by 100).

- The final table in the ANCOVA output (**Estimated marginal means**) provides us with the adjusted means on the dependent variable for each of our groups, split according to each of our independent variables separately and then jointly. 'Adjusted' refers to the fact that the effect of the covariate has been statistically removed. Given that the interaction effect was significant, our main focus is the final table labelled type of class* sex.

- As an optional extra I requested a plot of the adjusted means for the fear of statistics test, split for males and females and for the two interventions. (Don't worry if your graph does not look the same as mine: I modified my graph after it was generated so that it was easier to read when printed out. For instructions on modifying graphs, see Chapter 7.) It is clear to see from this plot that there is an interaction between the two independent variables. For males, Fear of Statistics scores were lowest (the aim of the program) in the maths skills intervention. For females, however, the lowest scores occurred in the confidence building intervention. This clearly suggests that males and females appear to respond differently to the programs, and that in designing interventions you must consider the gender of the participants.

Caution

Don't let yourself get too excited when you get a significant result. You must keep in mind what the analysis was all about. For example, the results here do not indicate that *all* males benefited from the maths skills program—nor that *all* females preferred the confidence building approach. Remember, we are comparing the mean score for the group as a whole. By summarising across the group as a whole we inevitably lose some information about individuals. In evaluating programs you may wish to consider additional analyses to explore how many people benefited versus how many people 'got worse'.

While one hopes that interventions designed to 'help' people will not do participants any harm, sometimes this does happen. In the field of stress management there have been studies that have shown an unexpected increase in participants' levels of anxiety after some types of treatment (e.g. relaxation training). The goal in this situation is to find out why this is the case for some people and to identify what the additional moderator variables are.

Presenting the results from two-way ANCOVA

The results of this analysis could be presented as follows:

A 2 by 2 between-groups analysis of covariance was conducted to assess the effectiveness of two programs in reducing fear of statistics for male and female participants. The independent variables were the type of program (maths skills, confidence building) and gender. The dependent variable was scores on the Fear of Statistics test (FOST), administered following completion of the intervention programs (Time 2). Scores on the FOST administered prior to the commencement of the programs (Time 1) were used as a covariate to control for individual differences.

Preliminary checks were conducted to ensure that there was no violation of the assumptions of normality, linearity, homogeneity of variances, homogeneity of regression slopes, and reliable measurement of the covariate. After adjusting for FOST scores at Time 1, there was a significant interaction effect [$F(1,25)=31.7$, $p<.0005$], with a large effect size (partial eta squared=.56). Neither of the main effects were statistically significant [program: $F(1,25)=1.43$, $p=.24$; gender: $F(1,25)=1.27$, $p=.27$]. These results suggest that males and females respond differently to the two types of interventions. Males showed a more substantial decrease in fear of statistics after participation in the maths skills program. Females, on the other hand, appeared to benefit more from the confidence building program.

In presenting the results of this analysis you would also provide a table of means for each of the groups. If the same scale is used to measure the dependent variable (Time 2) and the covariate (Time 1), then you would include the means at Time 1 and Time 2. You can get these easily by running **Descriptives** (see Chapter 6). If a different scale is used to measure the covariate you would provide the 'unadjusted' mean (and standard deviation) and the 'adjusted' mean (and standard error) for the two groups. The unadjusted mean is available from the Descriptive Statistics table. The adjusted mean (controlling for the covariate) is provided in the Estimated marginal means table. It is also a good idea to include the number of cases in each of your groups.

References

Cohen, J. (1988). *Statistical power analysis for the behavioral sciences*. Hillsdale, NJ: Erlbaum.

Stevens, J. (1996). *Applied multivariate statistics for the social sciences* (3rd edn). Mahway, NJ: Lawrence Erlbaum.

Tabachnick, B. G., & Fidell, L. S. (2001). *Using multivariate statistics* (4th edn). New York: HarperCollins.

22 Non-parametric statistics

In statistics books you will often see reference to two different types of statistical techniques: parametric and non-parametric. What is the difference between these two groups? Why is the distinction important? The word parametric comes from parameter, or characteristic of a population. The parametric tests (e.g. t-tests, analysis of variance) make assumptions about the population that the sample has been drawn from. This often includes assumptions about the shape of the population distribution (e.g. normally distributed). Non-parametric techniques, on the other hand, do not have such stringent requirements and do not make assumptions about the underlying population distribution (which is why they are sometimes referred to as distribution-free tests).

Despite being less 'fussy', non-parametric statistics do have their disadvantages. They tend to be less sensitive than their more powerful parametric cousins, and may therefore fail to detect differences between groups that actually exist. If you have the 'right' sort of data, it is always better to use a parametric technique if you can. So under what circumstances might you want or need to use non-parametric techniques?

Non-parametric techniques are ideal for use when you have data that are measured on nominal (categorical) and ordinal (ranked) scales. They are also useful when you have very small samples, and when your data do not meet the stringent assumptions of the parametric techniques. Although SPSS provides a wide variety of non-parametric techniques for a variety of situations, only the main ones will be discussed in this chapter. The topics covered in this chapter are presented below, along with their parametric alternative. Always check your data to see whether it is appropriate to use the parametric technique listed, in preference to using the less powerful non-parametric alternative.

Summary of techniques covered in this chapter

Non-parametric technique	Parametric alternative
Chi-square for independence	None
Mann-Whitney Test	Independent-samples t-test (Ch. 16)
Wilcoxon Signed Rank Test	Paired-samples t-test (Ch. 16)
Kruskal-Wallis Test	One-way between-groups ANOVA (Ch. 17)
Friedman Test	One-way repeated-measures ANOVA (Ch. 17)
Spearman Rank Order Correlation	Pearson's product-moment correlation (Ch. 11)

Assumptions for non-parametric techniques

Although the non-parametric techniques have less stringent assumptions, there are some general assumptions that should be checked.

- *Random samples.*
- *Independent observations.* Each person or case can be counted only once, they cannot appear in more than one category or group, and the data from one subject cannot influence the data from another. The exception to this is the repeated measures techniques (Wilcoxon Signed Rank Test, Friedman Test), where the same subjects are retested on different occasions or under different conditions.

Some of the techniques discussed in this chapter have additional assumptions that should be checked. These specific assumptions are discussed in the relevant sections.

Details of example

Throughout this chapter the various non-parametric techniques are illustrated using examples from the two data files on the website that accompanies this book (survey.sav, experim.sav; see p. xi). Full details of these data files are provided in the Appendix. The survey.sav file is used in the examples relating to chi-square, Mann-Whitney Test, Kruskal-Wallis Test and Spearman Rank Order Correlation. For techniques that involve a repeated measures design (Wilcoxon Signed Rank Test, Friedman Test), the experim.sav file is used. If you wish to follow along with the steps detailed in each of these examples, you will need to start SPSS and open the appropriate data file. These files can be opened only in SPSS.

This chapter provides only a brief summary of non-parametric techniques. For further reading, see the References at the end of this chapter.

Chi-square

There are two different types of chi-square tests, both involving categorical data:

1. The chi-square for goodness of fit (also referred to as one-sample chi-square) explores the proportion of cases that fall into the various categories of a *single* variable, and compares these with hypothesised values. This technique is not presented here but can be obtained from the **Nonparametric Tests** menu under the **Analyze** menu. For a review of chi-square see Chapter 17 in Gravetter and Wallnau (2000).
2. The chi-square test for independence is used to determine whether *two* categorical variables are related. It compares the frequency of cases found in the various categories of one variable across the different categories of another variable. For example: Is the proportion of smokers to non-smokers the same for males and females? Or, expressed another way: Are males more likely than females to be smokers?

Chi-square test for independence

This test is used when you wish to explore the relationship between *two* categorical variables. Each of these variables can have two or more categories. In the case where both of your variables have only two categories (resulting in a 2 by 2 table), some writers suggest that it may be more appropriate to use the phi-coefficient to describe the relationship between the variables. Phi can be requested using the same procedure as that used to obtain chi-square. Phi ranges from 0 to 1 and provides an indication of the strength of the relationship (in a similar manner to a correlation coefficient such as Pearson's r: see Chapter 11). For more information on phi, see any good statistics book (e.g. Gravetter & Wallnau, 2000, p. 605).

When a 2 by 2 table is encountered by SPSS the output from chi-square includes an additional correction value (Yates' Correction for Continuity). This is designed to correct or compensate for what some writers feel is an overestimate of the chi-square value when used with a 2 by 2 table.

In the following procedure I will demonstrate the use of chi-square using a 2 by 2 design; however, you should take into account the issues raised above. If your study involves variables with more than two categories (e.g. 2 by 3, 4 by 4), these issues won't concern you.

Summary for chi-square

Example of research question:	There are a variety of ways questions can be phrased: Are males more likely to be smokers than females? Is the proportion of males that smoke the same as the proportion of females? Is there a relationship between gender and smoking behaviour?
What you need:	Two categorical variables, with two or more categories in each, for example: • Gender (Male/Female); and • Smoker (Yes/No).
Additional assumptiopns (see general assumptions in the introduction to this chapter):	The lowest expected frequency in any cell should be 5 or more. Some authors suggest a less stringent criteria: at least 80 per cent of cells should have expected frequencies of 5 or more. If you have a 1 by 2 or a 2 by 2 table, it is recommended that the expected frequency be at least 10. If you have a 2 by 2 table that violates this assumption you should consider using Fisher's Exact Probability Test instead (also provided as part of the output from chi-square).
Parametric alternative:	none

Procedure for chi-square

1. From the menu at the top of the screen click on: **Analyze**, then click on **Descriptive Statistics**, then on **Crosstabs**.

2. Click on one of your variables (e.g. sex) to be your row variable, click on the arrow to move it into the box marked **Row(s)**.

3. Click on the other variable to be your column variable (e.g. smoker), click on the arrow to move it into the box marked **Column(s)**.

4. Click on the **Statistics** button. Choose **Chi-square**. Click on **Continue**.

5. Click on the **Cells** button.

6. In the **Counts** box, click on the **Observed** and **Expected** boxes.

7. In the **Percentage** section click on the **Row**, **Column** and **Total** boxes. Click on **Continue** and then **OK**.

The output generated from this procedure is shown below. It is for two variables each with two categories. If your variables have more than two categories the printout will look a little different, but the key information that you need to look for in the output is still the same. Only selected output is displayed.

SEX * SMOKE Crosstabulation

| | | | SMOKE | | |
			YES	NO	Total
SEX	MALES	Count	33	151	184
		Expected Count	35.9	148.1	184.0
		% within SEX	17.9%	82.1%	100.0%
		% within SMOKE	38.8%	43.0%	42.2%
		% of Total	7.6%	34.6%	42.2%
	FEMALES	Count	52	200	252
		Expected Count	49.1	202.9	252.0
		% within SEX	20.6%	79.4%	100.0%
		% within SMOKE	61.2%	57.0%	57.8%
		% of Total	11.9%	45.9%	57.8%
Total		Count	85	351	436
		Expected Count	85.0	351.0	436.0
		% within SEX	19.5%	80.5%	100.0%
		% within SMOKE	100.0%	100.0%	100.0%
		% of Total	19.5%	80.5%	100.0%

Chi-Square Tests

	Value	df	Asymp. Sig. (2-sided)	Exact Sig. (2-sided)	Exact Sig. (1-sided)
Pearson Chi-Square	.494[b]	1	.482		
Continuity Correction [a]	.337	1	.562		
Likelihood Ratio	.497	1	.481		
Fisher's Exact Test				.541	.282
Linear-by-Linear Association	.493	1	.483		
N of Valid Cases	436				

a. Computed only for a 2x2 table

b. 0 cells (.0%) have expected count less than 5. The minimum expected count is 35.87.

Interpretation of output from chi-square

The output from chi-square gives you a number of useful pieces of information.

Assumptions

The first thing you should check is whether you have violated one of the assumptions of chi-square concerning the 'minimum expected cell frequency', which should be 5 or greater (or at least 80 per cent of cells have expected frequencies of 5 or more). This information is given in a footnote below the final table (labelled Chi-Square Tests). Footnote b in the example provided indicates that '0 cells (.0%) have expected count less than 5'. This means that we have not violated the assumption, as all our expected cell sizes are greater than 5 (in our case greater than 35.87).

Chi-square tests

The main value that you are interested in from the output is the Pearson chi-square value, which is presented in the final table, headed **Chi-Square Tests**. If you have a 2 by 2 table (i.e. each variable has only two categories), then you should use the value in the second row (**Continuity Correction**). This is Yates' Correction for Continuity (which compensates for the overestimate of the chi-square value when used with a 2 by 2 table).

In the example presented above the corrected value is .337, with an associated significance level of .56 (this is presented in the column labelled **Asymp. Sig.** (2-sided). To be significant the Sig. value needs to be .05 or smaller. In this case the value of .56 is *larger* than the alpha value of .05, so we can conclude that our result is *not* significant. This means that the proportion of males that smoke is not significantly different from the proportion of females that smoke.

Summary information

To find what percentage of each sex smoke you will need to look at the summary information provided in the table labelled SEX*SMOKE Crosstabulation. This table may look a little confusing to start with, with a fair bit of information presented in each cell. To find out what percentage of males are smokers you need to read across the page in the first row, which refers to males. In this case we look at the values next to '% within sex'. For this example 17.9 per cent of males were smokers, while 82.1 per cent were non-smokers. For females, 20.6 per cent were smokers, 79.4 per cent non-smokers. If we wanted to know what percentage of the sample as a whole smoked we would move down to the total row, which summarises across both sexes. In this case we would look at the values next to '% of total'. According to these results, 19.5 per cent of the sample smoked, 80.5 per cent being non-smokers.

Mann-Whitney U Test

This technique is used to test for differences between two independent groups on a continuous measure. For example, do males and females differ in terms of their self-esteem? This test is the non-parametric alternative to the t-test for independent samples. Instead of comparing means of the two groups, as in the case of the t-test, the Mann-Whitney U Test actually compares medians. It converts the scores on the continuous variable to ranks, across the two groups. It then evaluates whether the ranks for the two groups differ significantly. As the scores are converted to ranks, the actual distribution of the scores does not matter.

Summary for Mann-Whitney U Test

Example of research question: Do males and females differ in terms of their levels of self-esteem?
    Do males have higher levels of self-esteem than females?
What do you need: Two variables:
  • one categorical variable with two groups (e.g. sex); and
  • one continuous variable (e.g. total self-esteem).
Assumptions: See general assumptions for non-parametric techniques presented at the beginning of this chapter.
Parametric alternative: Independent-samples t-test.

Procedure for Mann-Whitney U Test

1. From the menu at the top of the screen click on: **Analyze**, then click on **Non-parametric Tests**, then on **2 Independent Samples**.

2. Click on your continuous (dependent) variable (e.g. total self-esteem) and move it into the **Test Variable List** box.

3. Click on your categorical (independent) variable (e.g. sex) and move into **Grouping Variable** box.

4. Click on **Define Groups** button. Type in the value for Group 1 (e.g. 1) and for Group 2 (e.g. 2). These are the values that were used to code your values for this variable (see your codebook). Click on **Continue**.

5. Make sure that the **Mann-Whitney U** box is ticked under the section labelled **Test Type**. Click on **OK**.

The output generated from this procedure is shown below. Only selected output is displayed.

Test Statistics [a]

	Total self esteem
Mann-Whitney U	21594.000
Wilcoxon W	53472.000
Z	-1.227
Asymp. Sig. (2-tailed)	.220

[a]. Grouping Variable: SEX

Interpretation of output from Mann-Whitney U Test

The two values that you need to look at in your output are the Z value and the significance level, which is given as **Asymp. Sig (2-tailed)**. If your sample size is larger than 30, SPSS will give you the value for a Z-approximation test which includes a correction for ties in the data. In the example given above, the Z value is –1.23 (rounded) with a significance level of p=.22. The probability value (p) is not less than or equal to .05, so the result is not significant. There is no statistically significant difference in the self-esteem scores of males and females.

Wilcoxon Signed Rank Test

The Wilcoxon Signed Rank Test (also referred to as the Wilcoxon matched pairs signed ranks test) is designed for use with repeated measures: that is, when your subjects are measured on two occasions, or under two different conditions. It is the non-parametric alternative to the repeated measures t-test, but instead of comparing means the Wilcoxon converts scores to ranks and compares them at Time 1 and at Time 2. The Wilcoxon can also be used in situations involving a matched subject design, where subjects are matched on specific criteria.

To illustrate the use of this technique I have used data from the file labelled experim.sav included on the website accompanying this book (see p. xi, and the Appendix for details of the study). In this example I will compare the scores on a Fear of Statistics test, administered before and after an intervention designed to help students cope with a statistics course.

Summary for Wilcoxon Signed Rank Test

Example of research question: Is there a change in the scores on the Fear of Statistics test from Time 1 to Time 2?

What do you need: One group of subjects measured on the same continuous scale or criterion on two different occasions. The variables involved are scores at Time 1 or Condition 1, and scores at Time 2 or Condition 2.

Assumptions: See general assumptions for non-parametric techniques presented at the beginning of this chapter.

Parametric alternative: Paired-samples t-test.

Procedure for Wilcoxon Signed Rank Test

1. From the menu at the top of the screen click on: **Analyze**, then click on **Non-parametric Tests**, then on **2 Related Samples**.

2. Click on the variables that represent the scores at Time 1 and at Time 2 (e.g. fost1, fost2). Move these into the **Test Pairs List** box.

3. Make sure that the **Wilcoxon** box is ticked in the **Test Type** section. Click on **OK**.

The output generated from this procedure is shown below. Only selected output is displayed.

Test Statistics [b]

	fear of stats time2 - fear of stats time1
Z	-4.180[a]
Asymp. Sig. (2-tailed)	.000

a. Based on positive ranks.

b. Wilcoxon Signed Ranks Test

Interpretation of output from Wilcoxon Signed Rank Test

The two things you are interested in the output are the **Z** value and the associated significance levels, presented as **Asymp. Sig. (2-tailed)**. If the significance level is equal to or less than .05 (e.g. .04, .01, .001) then you can conclude that the difference between the two scores is statistically significant. In this example the Sig. value is .000 (which really means less than .0005). Therefore we can conclude that the two sets of scores are significantly different.

Kruskal-Wallis Test

The Kruskal-Wallis Test (sometimes referred to as the Kruskal-Wallis H Test) is the non-parametric alternative to a one-way between-groups analysis of variance. It allows you to compare the scores on some continuous variable for *three or more groups*. It is similar in nature to the Mann-Whitney test presented earlier in this chapter, but it allows you to compare more than just two groups. Scores are converted to ranks and the mean rank for each group is compared. This is a '*between-groups*' analysis, so different people must be in each of the different groups.

Summary for Kruskal-Wallis Test

Example of research question: Is there a difference in optimism levels across three age levels?

What you need: Two variables:
- one categorical independent variable with three or more categories (e.g. agegp3: 18–29, 30–44, 45+); and
- one continuous dependent variable (e.g. total optimism).

Assumptions: See general assumptions for non-parametric techniques presented at the beginning of this chapter.

Parametric alternative: One-way between-groups analysis of variance.

Procedure for Kruskal-Wallis Test

1. From the menu at the top of the screen click on: **Analyze**, then click on **Non-parametric Tests**, then on **K Independent Samples**.

2. Click on your continuous (dependent variable) (e.g. total optimism) and move it into the **Test Variable List** box.

3. Click on your categorical (independent variable) (e.g. agegp3) and move it into the **Grouping Variable** box.

4. Click on the **Define Range** button. Type in the first value of your categorical variable (e.g., 1) in the **Minimum** box. Type the largest value for your categorical variable (e.g. 3) in the **Maximum** box. Click on **Continue**.

5. In the **Test Type** section make sure that the **Kruskal-Wallis H** box is ticked. Click on **OK**.

The output generated from this procedure is shown below.

Ranks

	AGEGP	N	Mean Rank
Total Optimism	18-29	147	198.18
	30-44	153	216.05
	45+	135	241.80
	Total	435	

Test Statistics [a,b]

	Total Optimism
Chi-Square	8.573
df	2
Asymp. Sig.	.014

[a.] Kruskal Wallis Test

[b.] Grouping Variable : AGEGP3

Interpretation of output from Kruskal-Wallis Test

The main pieces of information you need from this output are: **Chi-Square** value, the degrees of freedom (**df**) and the significance level (presented as **Asymp. Sig.**). If this significance level is a value less than .05 (e.g. .04, .01, .001), then you can conclude that there is a statistically significant difference in your continuous variable across the three groups. You can then inspect the **Mean Rank** for the three groups presented in your first output table. This will tell you which of the groups had the highest overall ranking that corresponds to the highest score on your continuous variable.

In the output presented above the significance level was .01 (rounded). This is less than the alpha level of .05, so these results suggest that there is a difference in optimism levels across the different age groups. An inspection of the mean ranks for the groups suggest that the older group (45+) had the highest optimism scores, with the younger group reporting the lowest.

Friedman Test

The Friedman Test is the non-parametric alternative to the one-way repeated measures analysis of variance (see Chapter 17). It is used when you take the *same* sample of subjects or cases and you measure them at *three or more* points in time, or under three different conditions.

Summary for Friedman Test

Example of research question: Is there a change in Fear of Statistics scores across three time periods (pre-intervention, post-intervention and at follow-up)?

What do you need: One sample of subjects, measured on the same scale or measured at three different time periods, or under three different conditions.

Assumptions: See general assumptions for non-parametric techniques presented at the beginning of this chapter.

Parametric alternative: Repeated measures (within-subjects) analysis of variance.

Procedure for Friedman Test

1. From the menu at the top of the screen click on: **Analyze**, then click on **Non-parametric Tests**, then on **K Related Samples**.

2. Click on the variables that represent the three measurements (e.g. fost1, fost2, fost3).

3. In the **Test Type** section check that the **Friedman** option is selected. Click on **OK**.

The output generated from this procedure is shown below.

Ranks

	Mean Rank
fear of stats time 1	2.78
fear of stats time 2	2.03
fear of stats time 3	1.18

Test Statistics [a]

N	30
Chi-Square	41.568
df	2
Asymp. Sig.	.000

a. Friedman Test

Interpretation of output from Friedman Test

The results of this test suggest that there are significant differences in the Fear of Statistics scores across the three time periods. This is indicated by a Sig. level of .000 (which really means less than .0005). Comparing the ranks for the three sets of scores, it appears that there was a steady decrease in Fear of Statistics scores over time.

Spearman's Rank Order Correlation

Spearman's Rank Order Correlation (rho) is used to calculate the strength of the relationship between two continuous variables. This is the non-parametric alternative to Pearson's product-moment correlation. To illustrate the use of this technique I will use the same variables as were used when demonstrating Pearson's r (see Chapter 11). This analysis explores the relationship between perceived control (PCOISS) and perceived stress. The data file used in the example is survey.sav, which is included on the website with this book (see p. xi). The two variables are Total PCOISS and Total Perceived Stress (see the Appendix for details of the study).

Summary for Spearman's Rank Order Correlation

Example of research question:	How strong is the relationship between control of internal states (as measured by the PCOISS) and perceived stress (as measured by the Perceived Stress scale)?
What you need:	Two continuous variables (PCOISS, Total Perceived stress).
Assumptions:	See general assumptions for non-parametric techniques presented at the beginning of this chapter.
Parametric alternative:	Pearson's product-moment correlation coefficient.

Procedure for Spearman's Rank Order Correlation

The same basic procedure is used to request both Spearman correlation (rho) and Pearson's r. The only difference is the option that is ticked in the section labelled **Correlation coefficients**. If you wish you can even request both coefficients from the same analysis.

1. From the menu at the top of the screen click on: **Analyze**, then click on **Correlate**, then on **Bivariate**.

2 Click on your two variables (e.g. total PCOISS, total perceived stress) and move them into the box marked **Variables**.

3 In the section labelled **Correlation Coefficients** click on the option labelled **Spearman**. Click on **OK**.

The output generated from this procedure is shown below.

Correlations

			Total PCOISS	Total perceived stress
Spearman's rho	Total PCOISS	Correlation Coefficient	1.000	-.556**
		Sig. (2-tailed)	.	.000
		N	430	426
	Total perceived stress	Correlation Coefficient	-.556**	1.000
		Sig. (2-tailed)	.000	.
		N	426	433

** Correlation is significant at the .01 level (2-tailed).

Interpretation of output from Spearman's Rank Order Correlation

The output from Spearman Rank Order Correlation can be interpreted in the same way as the output obtained from Pearson product-moment correlation (see Chapter 11).

Additional exercises

Business

Data file: *staffsurvey.sav.* See Appendix for details of the data file.

1. Use Chi-square test for independence to compare the proportion of permanent versus casual staff (*employstatus*) who indicate they would recommend the organisation as a good place to work (*recommend*).
2. Use the Mann-Whitney U Test to compare the staff satisfaction scores (*totsatis*) for permanent and casual staff (*employstatus*).
3. Conduct a Kruskal-Wallis Test to compare staff satisfaction scores (*totsatis*) across each of the length of service categories (use the *servicegp3* variable).

Health

Data file: *sleep.sav.* See Appendix for details of the data file.

1. Use a Chi-square test for independence to compare the proportion of males and females (*gender*) who indicate they have a sleep problem (*problem*).

2. Use the Mann-Whitney U test to compare the mean sleepiness ratings (Sleepiness and Associated Sensations Scale total score: *totSAS*) for males and females (*gender*).

3. Conduct a Kruskal-Wallis Test to compare the mean sleepiness ratings (Sleepiness and Associated Sensations Scale total score: *totSAS*) for the three age groups defined by the variable *agegp3* (<=37, 38–50, 51+).

References

Daniel, W. (1990). *Applied nonparametric statistics* (2nd edn). Boston: PWS-Kent.

Gravetter, F. J., & Wallnau, L. B. (2000). *Statistics for the behavioral sciences* (5th edn). Belmont, CA: Wadsworth.

Siegel, S., & Castellan, N. (1988). *Nonparametric statistics for the behavioral sciences* (2nd edn). New York: McGraw-Hill.

Appendix Details of data files

This appendix contains information about the four data files that are included on the website accompanying this book (for details see p. xi):

- survey.sav
- experim.sav
- sleep.sav
- staffsurvey.sav

The first two files (*survey.sav* and *experim.sav*) are designed for you to follow along with the procedures described in the different chapters of this book. The second two files (*sleep.sav* and *staffsurvey.sav*) are additional files included to give you practice with data from different discipline areas. *Sleep.sav* includes data of a health and medical nature, while *staffsurvey.sav* contains data relevant to business and management. Exercises using these two files are suggested at the end of some of the chapters throughout the book.

To use the data files you will need to go to the website and download each file to your hard drive or to a floppy disk by following the instructions on screen. Then you should start SPSS and open the data file you wish to use. These files can only be opened in SPSS.

For each file, a codebook is included in this Appendix providing details of the variables and associated coding instructions.

Survey.sav

This is a real data file, condensed from a study that was conducted by my Graduate Diploma in Educational Psychology students. The study was designed to explore the factors that impact on respondents' psychological adjustment and wellbeing. The survey contained a variety of validated scales measuring constructs that the extensive literature on stress and coping suggest influence people's experience of stress. The scales measured self-esteem, optimism, perceptions of control, perceived stress, positive and negative affect, and life satisfaction. A scale was also included that measured people's tendency to present themselves in a favourable or socially desirable manner. The survey was distributed to members of the general public in Melbourne, Australia and surrounding districts. The final sample size was 439, consisting of 42 per cent males and 58 per cent females, with ages ranging from 18 to 82 (mean=37.4).

Experim.sav

This is a manufactured data set that was created to provide suitable data for the demonstration of statistical techniques such as t-test for repeated measures, and one-way ANOVA for repeated measures. This data set refers to a fictitious study that involves testing the impact of two different types of interventions in helping students cope with their anxiety concerning a forthcoming statistics course. Students were divided into two equal groups and asked to complete a number of scales (Time 1). These included a Fear of Statistics test, Confidence in Coping with Statistics scale and Depression scale.

One group (Group 1) was given a number of sessions designed to improve mathematical skills, the second group (Group 2) was subjected to a program designed to build confidence in the ability to cope with statistics. After the program (Time 2) they were again asked to complete the same scales that they completed before the program. They were also followed up three months later (Time 3). Their performance on a statistics exam was also measured.

Sleep.sav

This is real datafile condensed from a study conducted to explore the prevalence and impact of sleep problems on various aspects of people's lives. Staff from a university in Melbourne, Australia were invited to complete a questionnaire containing questions about their sleep behaviour (e.g. hours slept per night), sleep problems (e.g. difficulty getting to sleep) and the impact that these problems have on aspects of their lives (work, driving, relationships). The sample consisted of 271 respondents (55% female, 45% male) ranging in age from 18 to 84 years (mean=44yrs).

Staffsurvey.sav

This is a real datafile condensed from a study conducted to assess the satisfaction levels of staff from an educational institution with branches in a number of locations across Australia. Staff were asked to complete a short, anonymous questionnaire (shown later in this Appendix) containing questions about their opinion of various aspects of the organisation and the treatment they have received as employees.

Part A: Materials for survey.sav

Details of scales included in survey.sav

The scales are listed in the order in which they appear in the survey.

Scale	Reference
Life Orientation Test (Optimism) (*six items*)	Scheier, M. F., & Carver, C. S. (1985). Optimism, coping and health: An assessment and implications of generalized outcome expectancies. *Health Psychology, 4*, 219–247. Scheier, M. F., Carver, C. S. & Bridges, M. W. (1994). Distinguishing optimism from neuroticism (and trait anxiety, self-mastery and self-esteem): A re-evaluation of the Life Orientation Test. *Journal of Personality and Social Psychology, 67, 6*, 1063–1078.
Mastery Scale (*seven items*)	Pearlin, L., & Schooler, C. (1978). The structure of coping. *Journal of Health and Social Behavior, 19*, 2–21.
Positive and Negative Affect Scale (*twenty items*)	Watson, D., Clark, L. A. & Tellegen, A. (1988). Development and validation of brief measures of positive and negative affect: the PANAS scales. *Journal of Personality and Social Psychology, 54*, 1063–1070.
Satisfaction with Life Scale (*five items*)	Diener, E., Emmons, R. A., Larson, R. J. & Griffin, S. (1985). The Satisfaction with Life scale. *Journal of Personality Assessment, 49*, 71–76.
Perceived Stress Scale (*ten items*)	Cohen, S., Kamarck, T. & Mermelstein, R. (1983). A global measure of perceived stress. *Journal of Health and Social Behaviour, 24*, 385–396.
Self-esteem Scale (*ten items*)	Rosenberg, M. (1965). *Society and the adolescent self-image*. Princeton, NJ: Princeton University Press.
Social Desirability Scale (*ten items*)	Crowne, D. P., & Marlowe, P. (1960). A new scale of social desirability independent of psychochopathology. *Journal of Consulting Psychology, 24*, 349–354. Strahan, R., & Gerbasi, K. (1972). Short, homogeneous version of the Marlowe-Crowne Social Desirability Scale. *Journal of Clinical Psychology, 28*, 191–193.
Perceived Control of Internal States Scale (PCOISS) (*eighteen items*)	Pallant, J. (2000). Development and validation of a scale to measure perceived control of internal states. *Journal of Personality Assessment 75, 2*, 308–337.

Codebook for survey.sav

Full variable name	SPSS variable name	Coding instructions
Identification number	id	subject identification number
Sex	sex	1=males, 2=females
Age	age	in years
Marital	marital	1=single, 2=steady relationship, 3=living with a partner, 4=married for the first time, 5=remarried, 6=separated, 7=divorced, 8=widowed
Children	child	1=yes, 2=no
Highest level of education	educ	1=primary, 2=some secondary, 3=completed high school, 4=some additional training, 5=completed undergraduate, 6=completed postgraduate
Major source of stress	source	1=work, 2=spouse or partner, 3=relationships, 4=children, 5=family, 6=health/illness, 7=life in general
Do you smoke?	smoke	1=yes, 2=no
Cigarettes smoked per week	smokenum	Number of cigarettes smoked per week
Optimism scale	op1 to op6	1=strongly disagree, 5=strongly agree
Mastery scale	mast1 to mast7	1=strongly disagree, 4=strongly agree
PANAS scale	pn1 to pn20	1=very slightly, 5=extremely
Life Satisfaction scale	lifsat1 to lifsat5	1=strongly disagree, 7=strongly agree
Perceived Stress scale	pss1 to pss10	1=never, 5=very often
Self esteem scale	sest1 to sest10	1=strongly disagree, 4=strongly agree
Marlowe-Crowne Social Desirability scale	m1 to m10	1=true, 2=false
Perceived Control of Internal States scale (PCOISS)	pc1 to pc18	1=strongly disagree, 5=strongly agree

Total scale scores included in survey.sav

Full variable name	SPSS variable name	Coding instructions
Total Optimism	toptim	reverse items op2, op4, op6 add all scores op1 to op6 range 6 to 30
Total Mastery	tmast	reverse items mast1, mast3, mast4, mast6, mast7 add all items mast1 to mast7 range 7 to 28
Total Positive Affect	tposaff	add items pn1, pn4, pn6, pn7, pn9, pn12, pn13, pn15, pn17, pn18 range 10 to 50
Total Negative Affect	tnegaff	add items pn2, pn3, pn5, pn8, pn10, pn11, pn14, pn16, pn19, pn20 range 10 to 50
Total Life Satisfaction	tlifesat	add all items lifsat1 to lifsat5 range 5 to 35
Total Perceived Stress	tpstress	reverse items pss4, pss5, pss7, pss8 add all items pss1 to pss10 range 10 to 50
Total Self-esteem	tslfest	reverse items sest3, sest5, sest7, sest9, sest10 add all items sest1 to sest10 range 10 to 40
Total Social Desirability	tmarlow	reverse items m6 to m10 (recode true=1, false=0) add all items m1 to m10 range 0 to 10
Total Perceived Control of Internal States	tpcoiss	reverse items pc1, pc2, pc7, pc11, pc15, pc16 add all items pc1 to pc18 range 18 to 90
New Education categories	educ2	recoded the categories primary, some secondary into one group because of small numbers in each group: 1=primary/some secondary; 2=completed secondary; 3=some additional training; 4=completed undergraduate university; 5=completed postgraduate university
Age group 3 categories	agegp3	1=18–29yrs, 2=30–44yrs, 3=45+yrs
Age group 5 categories	agegp5	1=18–24yrs, 2=25–32yrs, 3=33–40yrs, 4=41–49yrs, 5=50+yrs

Copy of the questionnaire used in survey.sav

On the pages that follow, I have included a portion of the actual questionnaire used to collect the data included in the survey.sav file. The first page includes the demographic questions, followed by the Life Orientation Test (6 items) and the Positive and Negative Affect Scale (20 items). Full reference details of each scale included in the questionnaire are provided in the list given earlier in this appendix.

Sample questionnaire

1. Sex: ❏ male
❏ female *(please tick whichever applies)*

2. Age: _____ (in years)

3. What is your marital status? *(please tick whichever applies)*

   ❏ 1. single
   ❏ 2. in a steady relationship
   ❏ 3. living with partner
   ❏ 4. married for first time
   ❏ 5. remarried
   ❏ 6. separated
   ❏ 7. divorced
   ❏ 8. widowed

4. Do you have any children currently living at home with you?

   ❏ yes
   ❏ no *(please tick)*

5. What is the **highest** level of education that you have completed? *(please tick the **highest level** you have completed)*

   ❏ 1. primary school
   ❏ 2. some secondary school
   ❏ 3. completed secondary school
   ❏ 4. some additional training (apprenticeship, trade courses)
   ❏ 5. undergraduate university
   ❏ 6. postgraduate university

6. What are the major sources of stress in your life?

   _____

7. Do you smoke?

   ❏ yes
   ❏ no *(please tick)*

   If yes, how many cigarettes do you smoke per week? _____

Please read through the following statements and decide how much you either agree or disagree with each. Using the scale provided write the number that best indicates how you feel on the line next to each statement.

strongly disagree   1   2   3   4   5   **strongly agree**

1. _____ In uncertain times I usually expect the best.
2. _____ If something can go wrong for me it will.
3. _____ I'm always optimistic about my future.
4. _____ I hardly ever expect things to go my way.
5. _____ Overall I expect more good things to happen to me than bad.
6. _____ I rarely count on good things happening to me.

Source: Scheier, Carver & Bridges, 1994.

This scale consists of a number of words that describe different feelings and emotions. For each item indicate to what extent you have felt this way during the past few weeks. Write a number from 1 to 5 on the line next to each item.

very slightly or not at all	a little	moderately	quite a bit	extremely
1	2	3	4	5

1. _____ interested	8. _____ distressed	15. _____ excited
2. _____ upset	9. _____ strong	16. _____ guilty
3. _____ scared	10. _____ hostile	17. _____ enthusiastic
4. _____ proud	11. _____ irritable	18. _____ alert
5. _____ ashamed	12. _____ inspired	19. _____ nervous
6. _____ determined	13. _____ attentive	20. _____ jittery
7. _____ active	14. _____ afraid	

Source: Watson, Clark & Tellegen, 1988.

Part B: Materials for experim.sav

Codebook for experim.sav

Full variable name	SPSS variable name	SPSS variable label	Coding instructions
Id	id	id	Identification number
sex	sex	sex	1=males, 2=females
Age	age	age	in years
Group	group	type of class	1=maths skills, 2=confidence building
Fear of Statistics test at Time 1	fost1	fear of stats time1	Fear of Statistics test score at Time 1. Possible range 20–60. High scores indicate high levels of fear.
Confidence in Coping with Statistics Time1	conf1	confidence time1	Confidence in Coping with Statistics Test score at Time 1. Possible range 10–40. High scores indicate higher levels of confidence.
Depression Time 1	depress1	depression time1	Depression scale scores at Time 1. Possible range 20–60. High scores indicate high levels of depression.
Fear of Statistics test at Time 2	fost2	fear of stats time2	Fear of Statistics test score at Time 2. Possible range 20–60. High scores indicate high levels of fear.
Confidence in Coping with Statistics Time2	confid2	confidence time2	Confidence in Coping with Statistics test score at Time 2. Possible range 10–40. High scores indicate high levels of confidence.
Depression Time 2	depress2	depression time2	Depression scale scores at Time 2. Possible range 20–60. High scores indicate high levels of depression.
Fear of Statistics test at Time 3	fost3	fear of stats time3	Fear of Statistics test score at Time 3. Possible range 20–60. High scores indicate high levels of fear.
Confidence in Coping with Statistics Time3	conf3	confidence time3	Confidence in Coping with Statistics test score at Time 3. Possible range 10–40. High scores indicate high levels of confidence.
Depression Time 3	depress3	depression time3	Depression scale scores at Time 3. Possible range 20–60. High scores indicate high levels of depression.
Statistics Exam scores	exam	exam	Scores on the statistics exam. Possible range 0–100.

Part C: Materials for staffsurvey.sav

Staff survey (selected items)

Age: ❑ under 20
        ❑ 21 to 30
        ❑ 31 to 40
        ❑ 41 to 50
        ❑ over 50 yrs

Length of service with the organisation (in years): _____

Employment status: ❑ permanent ❑ casual

For each of the aspects shown below please rate your level of agreement and importance using the following scales:

Agreement: 1=not at all, 2=slight extent, 3=moderate extent, 4=great extent, 5=very great extent

Importance: 1=not at all, 2=slightly important, 3=moderately important, 4=very important, 5=extremely important

	Agreement	Importance
1. Is it clear what is expected of you at work?	1 2 3 4 5	1 2 3 4 5
2. At work have you been provided with all the equipment and materials required for you to do your work efficiently?	1 2 3 4 5	1 2 3 4 5
3. Does the organisation keep you up to date with information concerning development and changes?	1 2 3 4 5	1 2 3 4 5
4. Do you receive recognition from the organisation for doing good work?	1 2 3 4 5	1 2 3 4 5
5. Does your manager or supervisor encourage your development at work?	1 2 3 4 5	1 2 3 4 5
6. Do you feel that your opinions seem to count to the organisation?	1 2 3 4 5	1 2 3 4 5
7. Does the organisation make you feel that your job is important?	1 2 3 4 5	1 2 3 4 5
8. Do you feel that your fellow workers are committed to doing good quality work?	1 2 3 4 5	1 2 3 4 5
9. Has your performance been assessed or discussed in the last six months?	1 2 3 4 5	1 2 3 4 5
10. Have you had the opportunity over the last year at work to improve your skills?	1 2 3 4 5	1 2 3 4 5

Would you recommend this organisation as a good place to work?   ❑ Yes   ❑ No

Codebook for staffsurvey.sav

Description of variable	SPSS variable name	Coding instructions
Identification number	id	Subject identification number
City of residence of staff member	City	Each city was given a numerical code
Age of staff member	age	1=under 20, 2=21 to 30, 3=31 to 40, 4=41 to 50, 5=over 50
Years of service with the organisation	service	Years of service (if less than 1 record as decimal: e.g. 6mths=.5 year)
Employment status	employstatus	1=permanent, 2=casual
Q1 level of agreement	Q1a	1=not at all, 2=to a slight extent, 3=to a moderate extent, 4=to a great extent, 5=to a very great extent
Q1 level of importance	Q1imp	1=not important, 2=slightly important, 3=moderately important, 4=very important, 5=extremely important
Q2 level of agreement	Q2a	1=not at all, 2=to a slight extent, 3=to a moderate extent, 4=to a great extent, 5=to a very great extent
Q2 level of importance	Q2imp	1=not important, 2=slightly important, 3=moderately important, 4=very important, 5=extremely important
Q3 level of agreement	Q3a	1=not at all, 2=to a slight extent, 3=to a moderate extent, 4=to a great extent, 5=to a very great extent
Q3 level of importance	Q3imp	1=not important, 2=slightly important, 3=moderately important, 4=very important, 5=extremely important
Q4 level of agreement	Q4a	1=not at all, 2=to a slight extent, 3=to a moderate extent, 4=to a great extent, 5=to a very great extent
Q4 level of importance	Q4imp	1=not important, 2=slightly important, 3=moderately important, 4=very important, 5=extremely important
Q5 level of agreement	Q5a	1=not at all, 2=to a slight extent, 3=to a moderate extent, 4=to a great extent, 5=to a very great extent
Q5 level of importance	Q5imp	1=not important, 2=slightly important, 3=moderately important, 4=very important, 5=extremely important
Q6 level of agreement	Q6a	1=not at all, 2=to a slight extent, 3=to a moderate extent, 4=to a great extent, 5=to a very great extent
Q6 level of importance	Q6imp	1=not important, 2=slightly important, 3=moderately important, 4=very important, 5=extremely important
Q7 level of agreement	Q7a	1=not at all, 2=to a slight extent, 3=to a moderate extent, 4=to a great extent, 5=to a very great extent

Description of variable	SPSS variable name	Coding instructions
Q7 level of importance	Q7imp	1=not important, 2=slightly important, 3=moderately important, 4=very important, 5=extremely important
Q8 level of agreement	Q8a	1=not at all, 2=to a slight extent, 3=to a moderate extent, 4=to a great extent, 5=to a very great extent
Q8 level of importance	Q8imp	1=not important, 2=slightly important, 3=moderately important, 4=very important, 5=extremely important
Q9 level of agreement	Q9a	1=not at all, 2=to a slight extent, 3=to a moderate extent, 4=to a great extent, 5=to a very great extent
Q9 level of importance	Q9imp	1=not important, 2=slightly important, 3=moderately important, 4=very important, 5=extremely important
Q10 level of agreement	Q10a	1=not at all, 2=to a slight extent, 3=to a moderate extent, 4=to a great extent, 5=to a very great extent
Q10 level of importance	Q10imp	1=not important, 2=slightly important, 3=moderately important, 4=very important, 5=extremely important
Recommendation	recommend	0=no, 1=yes

Calculated variables

Total Staff Satisfaction Scale	totsatis	High scores indicate greater satisfaction
Length of service (3 groups)	Servicegp3	1=< or = to 2yrs, 2=3 to 5yrs, 3=6+ yrs
Log transformation of service	logservice	
Age recoded in 4 gps	agerecode	1=18 to 30yrs, 2=31 to 40 yrs, 3=41 to 50yrs, 4=50+

Part D: Materials for sleep.sav

Codebook for sleep.sav

Description of variable	SPSS variable name	Coding instructions
Identification Number	id	
Sex	sex	0=female, 1=male
Age	age	in years
Marital status	marital	1=single, 2=married/defacto, 3=divorced, 4=widowed
Highest education level achieved	edlevel	1=primary, 2=secondary, 3=trade, 4=undergrad, 5=postgrad
Weight (kg)	weight	in kg
Height (cm)	height	in cm
Rate general health	healthrate	1=very poor, 10=very good
Rate physical fitness	fitrate	1=very poor, 10=very good
Rate current weight	weightrate	1=very underweight, 10=very overweight
Do you smoke?	smoke	1=yes, 2=no
How many cigarettes per day?	smokenum	Cigs per day
How many alcoholic drinks per day?	alchohol	Drinks per day
How many caffeine drinks per day?	caffeine	Drinks per day
Hours sleep/weeknights	hourwnit	Hrs sleep on average each weeknight
Hours sleep/weekends	hourwend	Hrs sleep on average each weekend night
How many hours sleep needed?	hourneed	Hrs of sleep needed to not feel sleepy
Trouble falling asleep?	trubslep	1=yes, 2=no
Trouble staying asleep?	trubstay	1=yes, 2=no
Wake up during night?	wakenite	1=yes, 2=no
Work night shift?	niteshft	1=yes, 2=no
Light sleeper?	liteslp	1=yes, 2=no
Wake up feeling refreshed weekdays?	refreshd	1=yes, 2=no
Satisfaction with amount of sleep?	satsleep	1=very dissatisfied, 10=to a great extent
Rate quality of sleep	qualslp	1=very poor, 2=poor, 3=fair 4=good, 5=very good, 6=excellent
Rating of stress over last month	stressmo	1=not at all, 10=extremely
Medication to help you sleep?	medhelp	1=yes, 2=no
Do you have a problem with your sleep?	problem	1=yes, 2=no

Description of variable	SPSS variable name	Coding instructions
Rate impact of sleep problem on mood	impact1	1=not at all, 10=to a great extent
Rate impact of sleep problem on energy level	impact2	1=not at all, 10=to a great extent
Rate impact of sleep problem on concentration	impact3	1=not at all, 10=to a great extent
Rate impact of sleep problem on memory	impact4	1=not at all, 10=to a great extent
Rate impact of sleep problem on life sat	impact5	1=not at all, 10=to a great extent
Rate impact of sleep problem on overall wellbeing	impact6	1=not at all, 10=to a great extent
Rate impact of sleep problem on relationships	impact7	1=not at all, 10=to a great extent
Stop breathing during your sleep?	stopb	1=yes, 2=no
Restless sleeper?	restlss	1=yes, 2=no
Ever fallen asleep while driving?	drvsleep	1=yes, 2=no
Epworth sleepiness scale	ess	Total ESS score (range from 0=low to 24=high daytime sleepiness)
HADS Anxiety	anxiety	Total HADS Anxiety score (range from 0=no anxiety to 21=severe anxiety)
HADS Depression	depress	Total HADS Depression score (range from 0=no depression to 21=severe depression)
Rate level of fatigue over last week	fatigue	1=not at all, 10=to a great extent
Rate level of lethargy over last week	lethargy	1=not at all, 10=to a great extent
Rate how tired over last week	tired	1=not at all, 10=to a great extent
Rate how sleepy over last week	sleepy	1=not at all, 10=to a great extent
Rate lack energy over the last week	energy	1=not at all, 10=to a great extent
Problem staying asleep recoded	stayslprec	0=no, 1=yes
Problem getting to sleep recoded	getsleprec	0=no, 1=yes
Quality of sleep recoded into 4 groups	qualsleeprec	1=very poor, poor; 2=fair; 3=good; 4=very good, excellent
Sleepiness and Associated Sensations scale	totsas	Total Sleepiness and Associated Sensation Scale score (5=low, 50=extreme sleepiness)
Number of cigs per day recoded into 3 groups	cigsgp3	1=<=5, 2=6–15, 3=16+
Age recoded into 3 groups	agegp3	1=<=37yrs, 2=38–50yrs, 3=51+yrs
Problem with sleep recoded into 0/1	probsleeprec	0=no, 1=yes

Recommended references

Some of the articles and books I have found most useful for my own research and my teaching are listed here. Keep an eye out for new editions of these titles; many are updated every few years. I have classified these according to different headings, but many cover a variety of topics. The titles that I highly recommend have an asterisk next to them.

Research design

Bowling, A. (1997). *Research methods in health: Investigating health and health services*. Buckingham: Open University Press.
*Boyce, J. (2003). *Market research in practice*. Boston: McGraw-Hill.
*Cone, J., & Foster, S. (1993). *Dissertations and theses from start to finish*. Washington: American Psychological Association.
Goodwin, C. J. (1998). *Research in psychology: Methods and design* (2nd edn). New York: John Wiley.
Stangor, C. (1998). *Research methods for the behavioral sciences*. Boston: Houghton Mifflin.

Questionnaire design

*Oppenheim, A. N. (1992). *Questionnaire design, interviewing and attitude measurement*. London: St Martins Press.

Scale selection and construction

Dawis, R. V. (1987). Scale construction. *Journal of Counseling Psychology, 34*, 481–489.
DeVellis, R. F. (2003). *Scale development: Theory and applications* (2nd edn). Thousand Oaks, California: Sage.
Gable, R. K., & Wolf, M. B. (1993). *Instrument development in the affective domain: Measuring attitudes and values in corporate and school settings*. Boston: Kluner Academic.
Kline, P. (1986). *A handbook of test construction*. New York: Methuen.
*Robinson, J. P., Shaver, P. R., & Wrightsman, L. S. (Eds.). *Measures of personality and social psychological attitudes*. Hillsdale, NJ: Academic Press.
Streiner, D. L., & Norman, G. R. (1995). *Health measurement scales: A practical guide to their development and use* (2nd edn). Oxford: Oxford University Press.

Basic statistics

Cooper, D. R., & Schindler, P. S. (2003). *Business research methods* (8th edn). Boston: McGraw-Hill.

Everitt, B. S. (1996). *Making sense of statistics in psychology: A second level course.* Oxford: Oxford University Press.

*Gravetter, F. J., & Wallnau, L. B. (2000). *Statistics for the behavioral sciences* (5th edn). Belmont, CA: Wadsworth.

Norman, G. R., & Streiner, D. L. (2000). *Biostatistics: The bare essentials* (2nd edn). Hamilton: B.C. Decker Inc.

Pagano, R. R. (1998). *Understanding statistics in the behavioral sciences* (5th edn). Pacific Grove, CA: Brooks/Cole.

*Peat, J. (2001). *Health science research: A handbook of quantitative methods.* Sydney: Allen & Unwin.

Raymondo, J. C. (1999). *Statistical analysis in the behavioral sciences.* Boston: McGraw-Hill College.

Runyon, R. P., Coleman, K. A., & Pittenger, D. J. (2000). *Fundamentals of Behavioral Statistics* (9th edn). Boston: McGraw-Hill.

Smithson, M. (2000). *Statistics with confidence.* London: Sage.

Advanced statistics

Hair, J. F., Tatham, R. L., Anderson, R. E., & Black, W. C. (1998). *Multivariate data analysis* (5th edn). New York: Prentice Hall.

Stevens, J. (1996). *Applied multivariate statistics for the social sciences* (3rd edn). Mahway, NJ: Lawrence Erlbaum.

*Tabachnick, B. G., & Fidell, L. S. (2001). *Using multivariate statistics* (4th edn). New York: HarperCollins.

Preparing your report

American Psychological Association (2001). *Publication Manual* (5th edn). Washington: American Psychological Association.

Other references

Aiken, L. S., & West, S. G. (1991). *Multiple regression: Testing and interpreting interactions.* Newbury Park, CA: Sage.

Berry, W. D. (1993). *Understanding regression assumptions.* Newbury Park, CA: Sage.

Cohen, J. (1988). *Statistical power analysis for the behavioral sciences.* Hillsdale, NJ: Erlbaum.

Cohen, J., & Cohen, P. (1983). *Applied multiple regression/correlation analysis for the behavioral sciences* (2nd edn). New York: Erlbaum.

Cohen, S., Kamarck, T., & Mermelstein, R. (1983). A global measure of perceived stress. *Journal of Health and Social Behaviour,* 24, 385–396.

Crowne, D. P., & Marlowe, D. (1960). A new scale of social desirability independent of psychopathology. *Journal of Consulting Psychology, 24,* 349–354.

Diener, E., Emmons, R. A., Larson, R. J., & Griffin, S. (1985). The Satisfaction with Life scale. *Journal of Personality Assessment, 49,* 71–76.

Fox, J. (1991). *Regression diagnostics.* Newbury Park, CA: Sage.

Glass, G. V., Peckham, P. D., & Sanders, J. R. (1972). Consequences of failure to meet the assumptions underlying the use of analysis of variance and covariance. *Review of Educational Research, 42,* 237–288.

Greene, J., & d'Oliveira, M. (1999). *Learning to use statistical tests in psychology* (2nd edn). Buckingham: Open University Press.

Harris, R. J. (1994). *ANOVA: An analysis of variance primer.* Itasca, Ill: Peacock.

Hayes, N. (2000). *Doing psychological research: Gathering and analysing data.* Buckingham: Open University Press.

Keppel, G., & Zedeck, S. (1989). *Data analysis for research designs: Analysis of variance and multiple regression/correlation approaches.* New York: Freeman.

Pallant, J. (2000). Development and validation of a scale to measure perceived control of internal states. *Journal of Personality Assessment, 75,* 2, 308–337.

Pavot, W., Diener, E., Colvin, C. R., & Sandvik, E. (1991). Further validation of the Satisfaction with Life scale: Evidence for the cross method convergence of wellbeing measures. *Journal of Personality Assessment, 57,* 149–161.

Pearlin, L., & Schooler, C. (1978). The structure of coping. *Journal of Health and Social Behavior, 19,* 2–21.

Rosenberg, M. (1965). *Society and the adolescent self-image.* Princeton, NJ: Princeton University Press.

Scheier, M. F., & Carver, C. S. (1985). Optimism, coping and health: An assessment and implications of generalized outcome expectancies. *Health Psychology, 4,* 219–247.

Scheier, M. F., Carver, C. S., & Bridges, M. W. (1994). Distinguishing optimism from neuroticism (and trait anxiety, self-master, and self-esteem): A re-evaluation of the Life Orientation Test. *Journal of Personality and Social Psychology, 67,* 6, 1063–1078.

Strahan, R., & Gerbasi, K. (1972). Short, homogeneous version of the Marlowe-Crowne Social Desirability Scale. *Journal of Clinical Psychology, 28,* 191–193.

Watson, D., Clark, L. A., & Tellegen, A. (1988). Development and validation of brief measures of positive and negative affect: The PANAS scales. *Journal of Personality and Social Psychology, 54,* 1063–1070.

Index

Note: All entries in **bold** are SPSS procedures or SPSS dialogue boxes.